Advanced Information and Knowledge Processing

Series Editors
Professor Lakhmi Jain
Lakhmi.jain@unisa.edu.au
Professor Xindong Wu
xwu@cs.uvm.edu

For other titles published in this series, go to
www.springer.com/series/4738

Chenyi Hu · Ralph Baker Kearfott
André de Korvin · Vladik Kreinovich
Editors

Knowledge Processing with Interval and Soft Computing

Springer

Editors
Chenyi Hu
Computer Science Department
University of Central Arkansas
Conway, Arkansas, USA

Ralph Baker Kearfott
Mathematics Department
University of Louisiana, USA

André de Korvin
Computer and Mathematical Sciences Department
University of Houston-Downtown, USA

Vladik Kreinovich
Computer Science Department
University of Texas, USA

AI&KP ISSN 1610-3947
ISBN: 978-1-84996-784-6 e-ISBN: 978-1-84800-326-2
DOI: 10.1007/978-1-84800-326-2

British Library Cataloguing in Publication Data
A catalogue record for this book is available from the British Library

Printed on acid-free paper

Springer Science+Business Media
springer.com

Preface

Modern technologies have collected massive datasets from observations, experiments, and scientific simulation. Although progress has been made, it still remains a challenge to effectively and efficiently discover knowledge from such massive datasets. This is mainly because of the following two features. One is that the size and number of attributes (dimensions) of some datasets can be unmanageable. The other is that, due to the dynamic nature of the real world, changes and uncertainties are characteristics of datasets.

> A significant change in scientists' ability to analyze data to obtain a better understanding of natural phenomena will be enabled by (i) new ways to manage massive amounts of data from observations and scientific simulation, (ii) integration of powerful analysis tools directly into the database,
>
> ...
>
> Final report of the International Science 2020 Group, Microsoft, 2006

In contrast to classical point methods for arranging and processing data, in this book, we investigate strategies for knowledge processing with interval and soft computing.

Knowledge processing with interval methods has intrinsic merit. First, qualitative properties are often presented as ranges of data attributes rather than specific points. For example, one's blood pressure is normal if within the normal range (i. e. normal interval). By grouping attribute values into meaningful intervals, we can omit insignificant quantitative differences and focus more on qualitatively processing datasets. More importantly, interval-valued attributes contain more information than points and can represent variability and uncertainty. Finally, interval-valued computational results can be more meaningful and useful than point-valued output in a dynamic environment.

Statistical and probabilistic methods have been widely applied in knowledge discovery. However, despite the fact that confidence intervals and fuzzy intervals have been used to deal with uncertainties, they may not always work well in practice. By integrating interval methods with stochastic models and fuzzy logic, this book provides at least additional, if not more powerful, tools for knowledge processing, especially for handling variability and uncertainty.

Successful applications have been putting interval computing into the mainstream of computing. In 2006, the C++ standard committee evaluated a detailed proposal to include interval computing as a part of the ANSI/ISO C++ standard library. Interval arithmetic has already been in the kernel computations of Intel's Itanium-based architecture. Aside from many other software tools, Sun Microsystems has already included interval arithmetic in its Sun Studio.

More importantly, applying unique properties of interval computing, new algorithms have been developed to solve some otherwise very difficult problems. For example, one can computationally find all roots for nonlinear systems of equations on a given domain with interval Newton/generalized bisection methods and reliably find nonlinear global optima with interval branch-and-bound algorithms computationally. Very recently, Ferguson and Hales proved the 400-year-old Kepler conjecture with interval methods. In 2007, they received the first Robbins Prize from American Mathematical Society for their work.

Knowledge processing with intervals is significantly different from that with points. In this book, we extend previous knowledge processing methods to interval-valued datasets. By embedding interval and soft computing methods into distributed homogeneous and/or heterogeneous database systems that collect and manage massive datasets, scientists may significantly enhance their ability to process massive datasets.

This book can be used as an introduction to interval methods and soft computing for knowledge processing for upper-level undergraduates or first-year graduate students. It can also be a reference for researchers and practitioners.

We intended to make this book self-contained. Chapters 1, 2, and 3 provide necessary background knowledge for readers who are unfamiliar with interval and soft computing. Specifically, Chapter 1 introduces fundamentals of interval computing. In using interval computing for knowledge processing, soft computing technologies are applied. Therefore, Chapter 2 reviews essentials of soft computing. Although interval arithmetic and soft computing were developed independently, both of them can deal with uncertainty. We devote Chapter 3 to presenting their relationships. Readers familiar with these topics may skip the first three chapters.

Innovative algorithms and applications of interval and soft computing in knowledge processing are reported in Chapters 4 to 9. Specifically, Chapter 4 discusses knowledge processing methods related to interval linear algebra. Chapter 5 investigates interval function approximation. Chapter 6 presents an interval decision-making system. Chapter 7 studies interval-valued matrix games. Chapter 8 extends graph algorithms for interval-weighted graphs. Chapter 9 uses intervals in probabilistic studies. In Chapter 10, we present a standards-based object-oriented interval computing environment in C++. The entire software package is available at `http://www.cs.uca.edu/interval/`. Although these independent chapters cover different topics, there is some overlap. Each chapter is self-contained, but we reference other chapters as appropriate. By collecting our research results into a single volume, we unify and make accessible previously published work.

This book only introduces some initial applications of interval methods in knowledge processing. We sincerely hope to see more fruitful and significant results in both of theory and application in the future.

We would like to express our great appreciations to all co-authors of this volume. Hu, especially, would also like to acknowledge the US National Science Foundation for the grant awards of NSF/CISE/CCF-0202042 and NSF/CISE/CCF-0727798.

University of Central Arkansas, Conway, USA — *Chenyi Hu*
University of Louisiana, Lafayette, USA — *Ralph Baker Kearfott*
University of Houston-Downtown, Houston, USA — *André de Korvin*
University of Texas, El Paso, USA — *Vladik Kreinovich*

Contents

List of Contributors

Chenyi Hu
University of Central Arkansas
201 Donaghey Avenue
Conway, AR 72035, USA
chu@uca.edu

Ralph Baker Kearfott
University of Louisiana at Lafayette
Box 4-1010
Lafayette, LA 70504-1010, USA
rbk@louisiana.edu

Andre de Korvin
University of Houston-Downtown
One Main Street
Houston, TX 77002, USA
dekorvina@uhd.edu

Vladik Kreinovich
University of Texas at El Paso
500 W. University
El Paso, TX 79968, USA
vladik@utep.edu

Gary Anderson
University of Arkansas at Little Rock
2801 S. University Avenue
Little Rock, AR 72204, USA
gtanderson@ualr.edu

Daniel Berleant
University of Arkansas at Little Rock
2801 S. University Avenue
Little Rock, AR 72204, USA
jdberleant@ualr.edu

W. Dwayne Collins
Hendrix College
1600 Washington Avenue
Conway, AR 72032, USA
collins@hendrix.edu

Chaim Goodman-Strauss
University of Arkansas at Fayetteville
Fayetteville, AR 72701, USA
strauss@uark.edu

Ling T. He
University of Central Arkansas
201 Donaghey Avenue
Conway, AR 72035, USA
LingHe@uca.edu

Ping Hu
University of Central Arkansas
201 Donaghey Avenue
Conway, AR 72035, USA
PHu@uca.edu

Hong Lin
University of Houston-Downtown
One Main Street
Houston, TX 77002, USA
linh@uhd.edu

Michael Nooner
University of Central Arkansas
201 Donaghey Avenue
Conway, AR 72035, USA
MNooner@uca.edu

Plamen Simeonov
University of Houston-Downtown
One Main Street
Houston, TX 77002, USA
SimeonovP@uhd.edu

Shanying Xu
Academy of Mathematics and Systems Science
Chinese Academy of Sciences,
Beijing, China
xsy@iss.ac.cn

1

Fundamentals of Interval Computing

Ralph Baker Kearfott[1] and Chenyi Hu[2]

[1] Department of Mathematics, University of Louisiana at Lafayette, Box 4-1010, Lafayette, LA 70504-1010, USA. rbk@louisiana.edu

[2] Department of Computer Science, University of Central Arkansas, 201 Donaghey Avenue, Conway, AR 72035-0001, USA. chu@uca.edu

This volume deals, generally, with innovative techniques for automated knowledge representation and manipulation when such knowledge is subject to significant uncertainty, as well as with automated decision processes associated with such uncertain knowledge. Going beyond traditional probability theory and traditional statistical arguments, the techniques herein make use of interval techniques, of fuzzy knowledge representation and fuzzy logic, and of the combination of interval techniques with fuzzy logic and with probability theory.

In this chapter, we introduce interval computing, giving reasons for its development and references to historical work. We also preview the remainder of the book, contrasting the underlying philosophy and range of application with prevalent views among experts in interval computation.

1.1 Intervals and Their Representation

By the term *interval* we mean the set of all real numbers between specified lower and upper bounds, a and b (i. e. $\{x|a \leq x \leq b; a, b, x \in \mathbb{R}\}$). Intervals are denoted in various ways within the interval computations community. For examples, see [39, 18, 3, 40, 23]. In this book, we use lowercase boldface letters to denote intervals. For example, $\boldsymbol{x}$ is an interval. The lower and upper bounds of an interval $\boldsymbol{x}$ are specified as $\underline{x}$ and $\overline{x}$, respectively. Hence, $\boldsymbol{x} = [\underline{x}, \overline{x}]$. We call this the *endpoint representation* of an interval.[3] An empty interval - an interval that contains no real numbers - is simply the empty set $\emptyset$. Various computer representations are possible in implementations for $\emptyset$.

The *midpoint* of a nonempty interval is the algebraic average of its lower and upper bounds. The width of a nonempty interval is the difference between

[3] Alternate representations include midpoint-radius representation and tolerance representation. We discuss midpoint-radius (or midpoint-width) representation below.

C. Hu et al. (eds.), *Knowledge Processing with Interval and Soft Computing*,
DOI: 10.1007/978-1-84800-326-2_1,

its upper and lower bounds. We use the uppercase letters M and W in subscripts to specify the midpoint and width of a nonempty interval, respectively. The midpoint and width of the interval $\boldsymbol{x} = [\underline{x}, \overline{x}]$ are $x_M = (\underline{x} + \overline{x})/2$ and $x_W = \overline{x} - \underline{x}$, respectively. We define two unary interval operators, $m(\ )$ and $w(\ )$, that return the midpoint and width of an interval, respectively; that is, $m(\boldsymbol{x}) = x_M$ and $w(\boldsymbol{x}) = x_W$.

The lower bound of an interval is the same as the difference between its midpoint and one-half of its width. Similarly, the upper bound of an interval is the same as the sum of its midpoint and one-half of its width. We can represent an interval by its midpoint and width. This is called the *midpoint-width representation* of an interval. For example, the interval $\boldsymbol{x} = [\underline{x}, \overline{x}]$, in its midpoint-width representation, is $\boldsymbol{x} = [x_M - x_W/2, x_M + x_W/2]$. If the lower and upper bounds of an interval are the same, we say that is a *trivial* or *degenerate* or a *thin* interval. Obviously, the width of a degenerate interval is zero.

The above discussion can be extended to interval vectors and interval matrices. The entries of an interval vector or matrix are intervals. To unambiguously distinguish point vectors and interval vectors, and point matrices and interval matrices, in this book we use boldface letters to specify intervals. We use boldface lowercase letters with arrows on top to denote interval vectors. For example, $\vec{\boldsymbol{x}}$ is an interval vector. Its lower and upper bounds are $\underline{\vec{x}}$ and $\overline{\vec{x}}$, respectively. We omit the top arrow when it would not cause confusion. For example, $\vec{\boldsymbol{x}}$ (or $\boldsymbol{x}$) with

$$\vec{\boldsymbol{x}} = \begin{pmatrix} [1,2] \\ [3,4] \end{pmatrix}$$

can denote an interval vector.

Uppercase letters are used to denote matrices. Boldface uppercase letters denote interval matrices. For example, $\boldsymbol{A}$ is an interval matrix, and its lower and upper bounds are $\underline{A}$ and $\overline{A}$, respectively. The midpoint and width of a nonempty interval vector (or matrix) are real vectors (or matrices). Therefore, $\boldsymbol{A}$ with

$$\boldsymbol{A} = \begin{pmatrix} [1,2] & [3,4] \\ [5,6] & [7,8] \end{pmatrix}$$

can denote an interval matrix, and its lower and upper bounds are point matrices

$$\underline{A} = \begin{pmatrix} 1 & 3 \\ 5 & 7 \end{pmatrix},$$

$$\overline{A} = \begin{pmatrix} 2 & 4 \\ 6 & 8 \end{pmatrix}.$$

The notation we adopt in this volume has been recommended (but not required) for the journal *Reliable Computing* and has been proposed as a voluntary standard (although not universally adopted).

1.2 Origins and Reason for Development

Interval computing, specifically interval arithmetic, began primarily as a means of automating error analysis during the process of computing floating point approximations to solutions of scientific problems. The basic idea is that the true value x of some quantity appearing in some computation is not known exactly, but it is bounded in some interval $\boldsymbol{x}$, $x \in \boldsymbol{x}$. In exact arithmetic, each operation[4] $\odot$, $\odot \in \{+, -, \times, \div, \text{etc.}\}$ is formally defined on these intervals in such a way that the result $\boldsymbol{x} \odot \boldsymbol{y}$ is equal to the set $x \odot y$ as x ranges over all values in $\boldsymbol{x}$ and y ranges over all values in $\boldsymbol{y}$; that is,

$$\boldsymbol{x} \odot \boldsymbol{y} = \{x \odot y \mid x \in \boldsymbol{x} \text{ and } y \in \boldsymbol{y}\}\,. \tag{1.1}$$

For example, let $\boldsymbol{x} = [1, 2]$ and $\boldsymbol{y} = [-3, -1]$. We then have $\boldsymbol{x}+\boldsymbol{y} = [-2, 1]$, $\boldsymbol{x} - \boldsymbol{y} = [2, 5]$, $\boldsymbol{x} * \boldsymbol{y} = [-6, -1]$, and $\boldsymbol{x} \div \boldsymbol{y} = [-2, -1/3]$.

Using the basic interval arithmetic operations, we can perform linear algebra operations such as interval vector dot products, interval matrix-vector multiplications, interval matrix-matrix operations, etc. We will discuss and apply these in later chapters.

In addition to binary arithmetic operations, logic and set operations can also be performed on intervals. For example, $[0, 2] \subset [1, 2]$ returns false, $[0, 2] \cup [1, 3] = [0, 3]$, and $[0, 2] \cap [1, 3] = [1, 2]$. With these additional features of interval computing, we develop new algorithms to solve problems that are hard to solve in classical point arithmetic. By extending interval operations further with fuzzy logic and probability theory, this book presents additional algorithms for knowledge processing, especially in handling uncertainty.

Most modern literature on interval arithmetic is traceable to [36], although there are earlier independent works, such as [53], that are often overlooked but contain many, if not most, of the developments in [36]; some of these early works are available at `http://www.cs.utep.edu/interval-cmp/early.html`.

There is extensive literature on interval computations. Classic introductions are the books [37], [38], and [2] and the latter's translation to English [3]. A numerical analysis textbook that introduces interval arithmetic in appropriate places is [41]. An introduction to interval analysis with numerous computational examples within the MATLAB environment, as well as discussion of the most successful applications, appears in [39]. An early, carefully written reference on interval arithmetic and its implementation on actual machines is [29].

---

[4] In actual implementations, an operation $\odot$ can include not only the four elementary arithmetic operations but also all of the usual preprogrammed functions, such as sin, exp, etc., that are available in compiler libraries for scientific programming languages.

1.3 Computer Implementation and Software

In practice, (1.1) is not achievable exactly in floating point arithmetic on computers. However, on modern computers (and, in particular, on those for which the IEEE 754 binary floating point standard is implemented), *directed rounding* can be used, so that instead of the exact value $\boldsymbol{x} \odot \boldsymbol{y}$ defined by (1.1), an interval $\boldsymbol{z}$ is computed such that

$$\{x \odot y \mid x \in \boldsymbol{x} \text{ and } y \in \boldsymbol{y}\} \subseteq \boldsymbol{z}$$

and such that the complement of $\{x \odot y \mid x \in \boldsymbol{x} \text{ and } y \in \boldsymbol{y}\}$ in $\boldsymbol{z}$ is very small (on the order of the roundoff unit). In this way, if intervals $\{\boldsymbol{x}_1, \ldots, \boldsymbol{x}_n\}$ are substituted for the variables $\{x_1, \ldots, \boldsymbol{x}_n\}$ in an expression $E(x_1, \ldots, x_n)$, and E is evaluated using interval arithmetic, then the interval result $\boldsymbol{E}(\boldsymbol{x}_1, \ldots, \boldsymbol{x}_n)$ contains the range

$$\{E(x_1, \ldots, x_n) \mid x_i \in \boldsymbol{x}_i, \quad 1 \leq i \leq n\}, \tag{1.2}$$

that is, completion of a computation using interval arithmetic with directed rounding (also called *outwardly rounded interval arithmetic*), yielding a result $\boldsymbol{z}$, provides a mathematically rigorous proof[5] that the actual result is contained in $\boldsymbol{z}$.

There are numerous software tools and applications for interval computing written in mainstream languages, such as C, C++, Fortran, Java, Lisp, as well as in computational algebra systems, such as Maple, MATLAB, and Mathematica. For example, INTLAB [49] is an interval computing toolbox in MATLAB. An interval software development environment [22] is available in FORTRAN-90. Software in C++ supporting interval arithmetic includes the Boost C++ source libraries (`http://www.boost.org/libs/numeric/interval/doc/interval.htm`), filib++ [31], PROFIL/BIAS [26, 27], Sun's Studio C++, and Fortran [52], an object-oriented interval matrix computing environment [43], and others. Interval arithmetic is slated to be embodied in a technical report for the next C++ standards document [11]. We present an object-oriented interval toolbox in C++, named "IntBox," in Chapter 10 of this book.

Application software packages using interval computing include COSY-Infinity [9, 14] (and also `http://bt.pa.msu.edu/index_cosy.htm`) that is based on Taylor models and interval methods for validated solution of ordinary differential equations, quadrature, and range bounding, CGAL (`http://www.cgal.org/`) that makes geometric computations robust and efficient, GlobSol [23, 25] that finds reliable solutions for global nonlinear optimization problems with interval analysis, iCOs for interval constraint satisfaction and global optimization `http://ylebbah.googlepages.com/icos`, and others, both old and new. For additional applications, see [39].

[5] Assuming the computer is programmed correctly and is not malfunctioning.

1.4 Present Uses of Interval Arithmetic

People soon discovered that interval computations can be applied in more situations than merely error analysis in existing floating point algorithms. The fact that the result of a computation carried out with outwardly rounded interval arithmetic rigorously bounds the range of the computation is useful in various contexts. On a mathematical level, such uses include the following:

- rigorously bounding the ranges of functions over wide domains;
- including bounded uncertainties in the inputs in models;
- rigorously proving existence and uniqueness of solutions to systems of equations using computational versions of classical fixed point theorems.

As evidenced by the ubiquitous appearance of Lipschitz constants and moduli of continuity in classical hard analysis, bounding ranges of a quantity over sizable domains is an important computation in various contexts. Furthermore, using preexisting interval software and modern programming languages, bounding such ranges reduces to programming the computation of the quantity. Within this framework, computations that are equivalent to or sharper than using moduli of continuity or Lipschitz constants can be carried out automatically.

Classical fixed point theorems, such as the Brouwer fixed point theorem, state that if the image of a region $\boldsymbol{x}$ under an operator G is contained in $\boldsymbol{x}$, then there is an $x \in \boldsymbol{x}$ such that $G(x) = x$. With interval arithmetic, the range of G can be bounded, and the hypotheses of the theorem are satisfied if the interval evaluation of G over $\boldsymbol{x}$ is contained in $\boldsymbol{x}$. Furthermore, although such classical theorems can be applied directly, *interval Newton methods*, with a theoretical basis in classical fixed point theory, have been extensively developed to both compute narrow bounds on the solutions to systems of equations and to prove existence and uniqueness of those solutions within those bounds. For information on such methods, see, in addition to the general references on interval computations we have cited earlier, [40], [23], and [18]. The book [40] contains a thorough treatment of interval Newton methods, and [23] and [18] treat such methods in the context of algorithms for global optimization.

1.5 Pitfalls

Speed and sharpness were issues early in the study of interval methods, and there has been considerable discussion of these issues in the literature. However, present software implementations, using optimizing (in-lining) compilers and operator overloading, achieve, averaged over many operations, speed within a factor of 5 of hardware floating point operations, and some implementations within the compiler itself achieve interval evaluation that is, on average, less than a factor of 2 slower than floating point evaluations for some of the standard functions. Such speed is usually not the determining factor in whether to use interval computations.

Similarly, there has been significant discussion and development related to making the result intervals as tight as possible. In particular, many existing software systems today compute an interval $\boldsymbol{z}$ that is the narrowest machine-representable interval that contains the exact range as defined in (1.2) when $\odot \in \{+, -, \times, \div\}$; likewise, libraries for interval evaluation of the standard functions are also of high quality in this sense. Thus, tightness of the basic operations and interval evaluations of standard functions are not primary issues in deciding whether to apply interval computations.

The following pitfalls are more significant when designing applications with interval computations.

1.5.1 Interval Dependency

One major pitfall in interval computations is commonly termed *interval dependency*. Interval dependency is most easily illustrated by examining the subtraction operation: If $\boldsymbol{x} = [\underline{x}, \overline{x}]$ and $\boldsymbol{y} = [\underline{y}, \overline{y}]$, then the exact range $\boldsymbol{x} - \boldsymbol{y}$ happens to be

$$\boldsymbol{x} - \boldsymbol{y} = [\underline{x} - \overline{y}, \overline{x} - \underline{y}]. \tag{1.3}$$

However, suppose, say, $\boldsymbol{x} = [-1, 1]$ is interpreted to represent some specific real number $x \in [-1, 1]$ but unknown other than that it lies in $[-1, 1]$. Then $x - x = 0$, but if the computer encounters the expression $x - x$ (without simplification) and substitutes $[-1, 1]$ for both instances of x, it obtains

$$[-1, 1] - [-1, 1] = [-2, 2]. \tag{1.4}$$

Observe that $[-2, 2]$ does contain the range of $f(x) = x - x$ as x ranges over $[-1, 1]$, but it does not sharply bound the range. The "dependency" is from the fact that the computation (1.4) assumes implicitly that the quantity in the first instance of $[-1, 1]$ varies independently from the quantity in the second instance of $[-1, 1]$ whereas, in fact, the two values are correlated or "dependent." Observe, however, that $[-2, 2]$ *is* the exact range of $g(x, y) = x - y$ as x ranges over $[-1, 1]$ and y ranges over $[-1, 1]$.

Due partly to the interval dependency phenomenon, traditional floating point algorithms can seldom be converted to successful interval algorithms (that rigorously bound roundoff error or provide useful bounds on ranges) by simply replacing floating point numbers by intervals; instead, intervals need to be introduced in appropriate ways, and new algorithms appropriate for interval computation are developed.

One property of interval ranges that ameliorates interval dependency is that the amount of overestimation (i. e., the sum of the differences between the endpoints of the actual range and the interval arithmetic evaluation) decreases proportionally to the widths of the input intervals. In fact, for continuously differentiable quantities, the bounds on the range can be computed in such a way that the decrease is proportional to the squares of the widths of the input intervals; in special circumstances, bounds can even be computed with overestimation proportional to even higher powers of the widths of the input intervals. Thus, interval computations give locally tight bounds on the ranges of computed quantities.

1.5.2 Computational Complexity

Computational complexity and similar theoretical issues arise in interval algorithms. For instance, if $\boldsymbol{A}$ is a matrix with interval entries, $\boldsymbol{b}$ is a vector with interval entries, and the individual entries in $\boldsymbol{A}$ and $\boldsymbol{b}$ are assumed to vary independently, then it is known that, in general, finding an interval vector sharply bounding the solution set

$$\{x \mid Ax = b \text{ for some } A \in \boldsymbol{A} \text{ and } b \in \boldsymbol{b}\} \tag{1.5}$$

is an NP-hard problem. (Such results are collected in [28].) However, there are many algorithms that successfully compute usefully narrow bounds to the solution sets to even large linear systems; as one of many examples of success, see [42]. Furthermore, there is general software that is at least moderately successful at handling large, sparse systems of linear equations within a friendly user environment: the MATLAB toolbox INTLAB (see [48, 19]).

1.5.3 Problems with Coordinate Systems

Actual ranges of quantities as input values range over intervals are seldom sets of the form

$$\{(x_1, \ldots, x_n) \mid x_i \in [\underline{x}_i, \overline{x}_i] \text{ for } 1 \le i \le n\}\,. \tag{1.6}$$

(Such sets are commonly termed *boxes* or *interval vectors*.) For instance, the image of the vector-valued function $F(x,y) = (x+y, x-y)^T$ is the skewed parallelogram depicted in Figure 1.1, whereas the smallest interval vector containing this range is the substantially larger vertically oriented box in Figure 1.1. This problem can be severe in interval methods for bounding the solution sets to initial value problems for systems of ordinary differential equations, where it is known as the *wrapping effect* (see, for instance, [38]). It even is a significant problem in branch-and-bound methods for global optimization, where excessive subdivision of a domain may occur if the proper coordinate system is not used.

The wrapping effect has been, to a large extent, successfully ameliorated for initial value problems, such as in the COSY system [10]. In general, a fix is to somehow use an appropriate change of coordinates, as is proposed in [24].

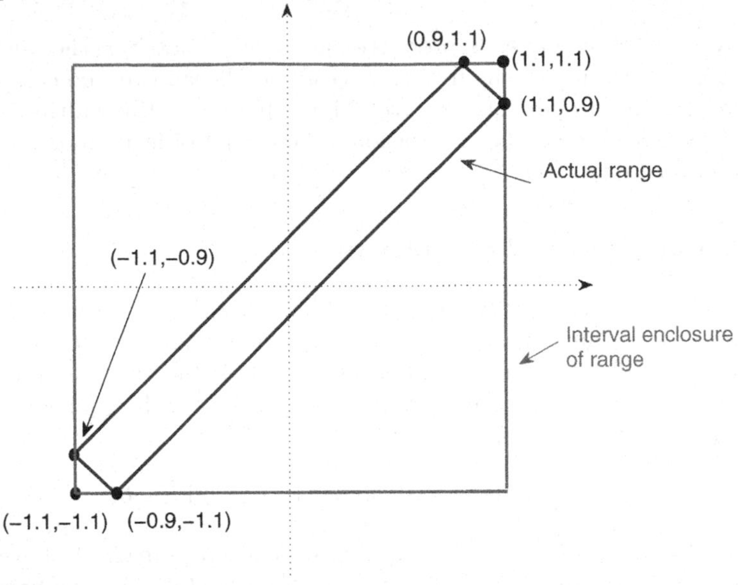

Fig. 1.1. Illustration of overestimation due to the choice of coordinates.

1.5.4 Successes Nonetheless

The concept of mathematical rigor in computer arithmetic is appealing, something that has caused enthusiasts to make excessive claims about its applicability and usefulness in the past. These claims, in turn, have stimulated naive experimentation, leading to disillusionment with the technology within the scientific computing community as a whole. However, with software crafted to utilize the strengths and avoid the weaknesses of interval computations, important problems are increasingly solved with interval techniques but not other methods. For instance, certain chemical kinetics equilibrium problems [50, 51, 15, 16] have been solved correctly with interval techniques, whereas earlier floating techniques had given approximate solutions that led to erroneous conclusions about the underlying physical problem. There have been similar successes in solving systems of equations arising in robotics [20, 30]. Inroads have even been made in the computation of parameters for boundary value problems [33, 32, 34, 35]. For additional details on these and other applications, see [39].

1.6 Context of This Work

In this volume, we present novel techniques for new and improved algorithms in knowledge engineering. In contrast to much work utilizing interval arithmetic, the focus here is not on rigor and mathematical theorem proving but on efficient ways of encompassing uncertain inputs to compute bounds on outputs. Along these lines, intervals are combined heuristically with methods from probability theory and other methods that, in their raw form, cannot be made rigorous by naive application of interval arithmetic. A guiding principle in the applications treated in this volume is that the real-world data often provide interval inputs to the problem and that the data should be so represented.

In some of the problems tackled in this volume, traditional statistical models have previously been used, but interval techniques are used creatively for improved modeling and predictive power. An example of this, appearing in Section 4.5 of Chapter 4, is the use of interval data to reduce the raw noisy data in stock market indices. This data reduction is combined with an innovative view[6] of interval values of the singular value decomposition and principal component analysis to both simplify and enhance the predictive power of the model. In this work and Chapter 5, intervals and interval arithmetic are used to describe variability and uncertainty in the model inputs and to divine relationships between the model inputs and model outputs; the model outputs are intervals that approximate the range of behavior over the input intervals but are not claimed to rigorously enclose that range.[7] This is reasonable in view of the fact that real-world models often have uncertainties that cannot be quantified, so proof that an exact result is rigorously enclosed may not make sense.

In contrast, the traditional literature on interval enclosures for eigenvalue- and singular-value-decompositions assumes that the problem is specified exactly (with a point matrix), and the goal is to compute mathematically rigorous bounds on the exact solutions to this point problem. Examples of this approach are in [21, 4, 5, 6, 7, 1], and [44, 45, 8, 54, 13] in applications to partial differential equations. Although there has been some work, such as [46, 47, 12, 13], on bounding the range of eigenvalues of an interval matrix, rigorous enclosure of the range (often called *outer enclosures*) is problematical for general matrices. For this reason, when we adopt a statistician's philosophy and employ intervals to reduce noisy data and obtain approximate bounds on ranges, we do not obtain mathematically rigorous results, but we may get reasonable "guesses" in problems that otherwise would be intractable or for which rigorous bounds are not meaningful. Furthermore, using interval technology, we obtain new approaches that compare favorably with traditional statistical methods.

Continued research on the methods in this volume should lead to additional mathematical rigor and explanations for the reasons the models appear to work so well.

---

[6] But similar to that in [17].

[7] Nonetheless, measures of quality of the interval result, in terms of how close it is to the exact range, are discussed.

References

1. Aberth, O., Schaefer, M.J.: Precise matrix eigenvalues using range arithmetic. SIAM Journal on Matrix Analysis and Applications 14(1), 235–241 (1993)
2. Alefeld, G., Herzberger, J.: Einführung in die Intervallrechnung. Springer-Verlag, Berlin, (1974)
3. Alefeld, G., Herzberger, J.: Introduction to Interval Computations. Academic Press Inc., New York (1983)
4. Alefeld, G.: Componentwise inclusion and exclusion sets for solutions of quadratic equations in finite-dimensional spaces. Numerische Mathematik 48(4), 391–416 (1986)
5. Alefeld, G., Spreuer, H.: Iterative improvement of componentwise errorbounds for invariant subspaces belonging to a double or nearly double eigenvalue. Computing 36, 321–334 (1986)
6. Behnke, H.: Inclusion of eigenvalues of general eigenvalue problems for matrices. In: U. Kulisch, H.J. Stetter (eds.) Scientific Computation with Automatic Result Verification, Computing. Supplementum, Vol. 6, pp. 69–78. Springer, New York (1988)
7. Behnke, H.: Bounds for eigenvalues of parameter-dependent matrices. Computing 49(2), 159–167 (1992)
8. Behnke, H., Mertins, U.: Bounds for eigenvalues with the use of finite elements. In: U. Kulisch, R. Lohner, A. Facius (eds.) Perspectives on Enclosure Methods: GAMM-IMACS International Symposium on Scientific Computing, Computer Arithmetic and Validated Numerics, September 2000, Karlsruhe, Germany, 119–132. Kluwer Academic Publishers, Amsterdam (2001)
9. Berz, M., Makino, K., Shamseddine, K., Hoffstätter, G.H., Wan, W.: COSY INFINITY and its applications to nonlinear dynamics. In: M. Berz, C. Bischof, G. Corliss, A. Griewank (eds.) Computational Differentiation: Techniques, Applications, and Tools, 363–365. SIAM, Philadelphia (1996)
10. Berz, M.: COSY INFINITY web page `http://cosy.pa.msu.edu/cosy.pa.msu.edu` (2000)
11. Brönnimann, H., Melquiond, G., Pion, S.: A proposal to add interval arithmetic to the C++ Standard Library. Technical proposal N1843-05-0103, CIS, Brooklyn Polytechnic University, S Brooklyn (2005)
12. Chen, S., Qiu, Z., Liu, Z.: A method for computing eigenvalue bounds in structural vibration systems with interval parameters. Computers and Structures 51(3), 309 (1994)
13. Chen, S., Qiu, Z., Liu, Z.: Perturbation method for computing eigenvalue bounds in structural vibration systems with interval parameters. Communications in Applied Numerical Methods 10(2), 121–134 (1994)
14. Corliss, G.F., Yu, J.: Testing COSY's interval and Taylor model arithmetic. In: R. Alt, A. Frommer, R.B. Kearfott, W. Luther (eds.) Numerical Software with Result Verification: Platforms, Algorithms, Applications in Engineering, Physics, and Economics, Lectures Notes in Computer Science, No. 2992, pp. 91–105. Springer, Heidelberg (2004)
15. Gau, C.Y., Stadtherr, M.A.: New interval methodologies for reliable chemical process modeling. Computers and Chemical Engineering 26, 827–840 (2002)
16. Gau, C.Y., Stadtherr, M.A.: Dynamic load balancing for parallel interval-Newton using message passing. Computers and Chemical Engineering 26, 811–815 (2002)

17. Gioia, F., Lauro, C.N.: Principal component analysis on interval data. Computational Statistics 21(2), 343–363 (2006)
18. Hansen, E.R., Walster, W.: Global Optimization Using Interval Analysis, 2nd ed. Marcel Dekker, New York (2003)
19. Hargreaves, G.I.: Interval analysis in MATLAB. Master's thesis, Department of Mathematics, University of Manchester (2002)
20. Jaulin, L., Keiffer, M., Didrit, O.,Walter, E.: Applied Interval Analysis. Springer-Verlag, Berlin (2001)
21. Kalmykov, S.A.: To the problem of determination of the symmetric matrix eigenvalues by means of the interval method. In: Numerical Analysis, Collect. Sci. Works, pp. 55–59. Sov. Acad. Sci., Sib. Branch, Inst. Theor. Appl. Mech., Novosibirsk, USSR (1978) (in Russian)
22. Kearfott, R.B.: A Fortran 90 environment for research and prototyping of enclosure algorithms for nonlinear equations and global optimization. ACM Transactions on Mathematical Software 21(1), 63–78 (1995)
23. Kearfott, R.B.: Rigorous Global Search: Continuous Problems. Nonconvex Optimization and Its Applications. No. 13. Kluwer Academic, Norwell, MA (1996)
24. Kearfott, R.B.: Verified branch and bound for singular linear and nonlinear programs: An epsilon-inflation process (April 2007), Submitted
25. Kearfott, R.B.: GlobSol User Guide. Optimization Methods and Software (2008). Submitted
26. Knüppel, O.: PROFIL/BIAS - A fast interval library. Computing 53(3–4), 277–287 (1994)
27. Knüppel, O.: PROFIL/BIAS v 2.0. Bericht 99.1, Technische Universität Hamburg-Harburg, Harburg, Germany (1999). Available from `http://www.ti3.tu-harburg.de/profil_e`
28. Kreinovich, V., Lakeyev, A., Rohn, J., Kahl, P.: Computational Complexity and Feasibility of Data Processing and Interval Computations, Applied Optimization, Vol. 10. Kluwer Academic, Norwell, MA (1998)
29. Kulisch, U.W., Miranker, W.L.: Computer Arithmetic in Theory and Practice. Computer Science and Applied Mathematics. Academic Press Inc., New York (1981)
30. Lee, D., Mavroidis, C., Merlet, J.P.: Five precision point synthesis of spatial RRR manipulators using interval analysis. Journal of Mechanical Design 126, 842–849 (2004)
31. Lerch, M., Tischler, G., Gudenberg, J.W.V., Hofschuster, W., Krämer, W.: FILIB++, a fast interval library supporting containment computations. ACM Transactions on Mathematical Software 32(2), 299–324 (2006)
32. Lin, Y., Stadtherr, M.A.: Advances in interval methods for deterministic global optimization in chemical engineering. Journal of Global Optimization 29, 281–296 (2004)
33. Lin, Y., Stadtherr, M.A.: Lp strategy for interval-Newton method in deterministic global optimization. Industrial & Engineering Chemistry Research, 43, 3741–3749 (2004)
34. Lin, Y., Stadtherr, M.A.: Locating stationary points of sorbate-zeolite potential energy surfaces using interval analysis. J. Chemical Physics, 121, 10159-10166 (2004)
35. Lin, Y., Stadtherr, M.A.: Deterministic global optimization of molecular structures using interval analysis. J. Computational Chemistry 26, 1413–1420 (2005)

36. Moore, R.E.: Interval arithmetic and automatic error analysis in digital computing. Ph.D. dissertation, Department of Mathematics, Stanford University, Stanford, CA (1962)
37. Moore, R.E.: Interval Analysis. Prentice–Hall, Upper Saddle River, NJ (1966)
38. Moore, R.E.: Methods and Applications of Interval Analysis. Society for Industrial and Applied Mathematics, Philadelphia (1979)
39. Moore, R.E., Kearfott, R.B., Cloud, M.J.: Introduction to interval algorithms and their applications with INTLAB: a MATLAB toolkit. Submitted
40. Neumaier, A.: Interval Methods for Systems of Equations. Encyclopedia of Mathematics and Its Applications, Vol. 37. Cambridge University Press, Cambridge (1990)
41. Neumaier, A.: Introduction to Numerical Analysis. Cambridge University Press, Cambridge (2001)
42. Neumaier, A., Pownuk, A.: Linear systems with large uncertainties, with applications to truss structures. Reliable Computing 13, 149–172 (2007)
43. Nooner, M., Hu, C.: A computational environment for interval matrices. In: R.L. Muhanna, R.L. Mullen (eds.) Proceedings of 2006 Workshop on Reliable Engineering Computing, pp. 65–74. Georgia Tech. University, Savanna (2006). `http://www.gtsav.gatech.edu/workshop/rec06/proceedings.html`
44. Oishi, S.: Fast enclosure of matrix eigenvalues and singular values via rounding mode controlled computation. Linear Algebra and its Applications 324(1–3), 133–146 (2001)
45. Plum, M.: Computer-assisted enclosure methods for elliptic differential equations. Linear Algebra and its Applications 324(1–3), 147–187 (2001)
46. Rohn, J., Deif, A.: On the range of eigenvalues of an interval matrix. Computing 47(3–4), 373–377 (1992)
47. Rohn, J.: Interval matrices: Singularity and real eigenvalues. SIAM Journal on Matrix Analysis and Applications 14(1), 82–91 (1993)
48. Rump, S.M.: INTLAB-INTerval LABoratory. In: T. Csendes (ed.) Developments in Reliable Computing: Papers presented at the International Symposium on Scientific Computing, Computer Arithmetic, and Validated Numerics, Vol. 5(3), pp. 77–104. Kluwer Academic, Norwell, MA (1999)
49. Rump, S.M.: INTLAB - INTerval LABoratory (1999-2008) `http://www.ti3.tu-harburg.de/rump/intlab/`
50. Stadtherr, M.A.: Interval analysis: Application to phase equilibrium problems. In: A. Iserles (ed.) Encyclopedia of Optimization. Kluwer Academic, Norwell, MA (2001)
51. Stadtherr, M.A.: Interval analysis: Application to chemical engineering design problems. In: A. Iserles (ed.) Encyclopedia of Optimization. Kluwer Academic, Norwell, MA (2001)
52. Sun: Sun studio math libraries (1994-2007). Available from `http://developers.sun.com/sunstudio/documentation/libraries/math_libraries.jsp`
53. Sunaga, T.: Theory of interval algebra and its application to numerical analysis. RAAG Memoirs 2, 29–46 (1958)
54. Wieners, C.: A parallel Newton multigrid method for high order finite elements and its application on numerical existence proofs for elliptic boundary value equation. Zeitschrift für Angewandte Mathematik und Mechanik 76, 171–176 (1996)

2

Soft Computing Essentials

Andre de Korvin, Hong Lin, and Plamen Simeonov

Department of Computer and Mathematical Sciences, University of Houston, Downtown, One Main Street, Houston, Texas 77002, USA. dekorvina@uhd.edu

The main purpose of this chapter is to give an overview of some of the soft computing methods that are currently applied to a variety of problems. One of the characteristics of soft computing methods is that they are typically used in problems where mathematical models are not available or are intractable or too cumbersome to be viable. Another characteristics is that uncertainty inherent in many situations under study is taken into account rather than ignored. Often a human expert has partial knowledge and it is this partial knowledge (as opposed to complete knowledge) that soft computing uses to advantage. Another characteristic is that soft computing often provides a good solution as opposed to an optimal solution. In this chapter we present some of the mainstream methods of soft computing: Neural Nets, Fuzzy Logic, Neuro-Fuzzy Systems, The Theory of Evidence, Rough Sets, and Genetic Algorithms.

2.1 Neural Nets

A large number of structures have been used and we will highlight a small sample of the great variety of networks used. For a small fraction of the material available for neural nets we refer the reader to [6], [8], [11], [5], [9], [23], and [39]. A neural net is a set of neurons. Each neuron is as shown in Figure 2.1.

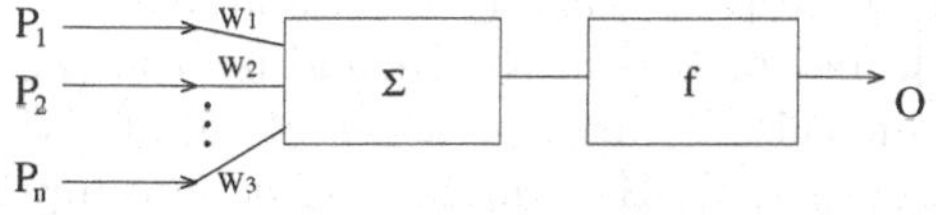

Fig. 2.1

An n-dimensional input $(p_1, \ldots, p_n)^T$ is provided. The first component of the neuron carries a weighted average called the net input:

C. Hu et al. (eds.), *Knowledge Processing with Interval and Soft Computing*, DOI: 10.1007/978-1-84800-326-2_2,

$$N = \sum_{i=1}^{n} w_i p_i.$$

The second neuron transforms N into the output

$$O = f(N).$$

We refer to f as the transfer function. Commonly used transfer functions are

$$\begin{aligned} f(N) = \text{Hardlim}(N) &= \begin{cases} 1 \text{ if } N \geq 0 \\ 0 \text{ otherwise,} \end{cases} \\ f(N) = \text{Hardlims}(N) &= \begin{cases} 1 \text{ if } N \geq 0 \\ -1 \text{ otherwise,} \end{cases} \\ f(N) = \text{LogSig}(N) &= 1/(1 + e^{-N}). \end{aligned}$$

Often an additional input of 1 is provided as well as an additional weight called the bias. The net input is then given by

$$N = \sum_{i=1}^{n} w_i p_i + b,$$

where b is the bias. The set of neurons forming a neural net is often organized into layers as shown for example in Figure 2.2.

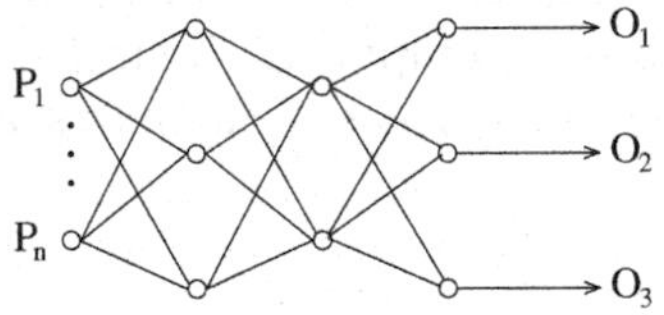

Fig. 2.2

In Figure 2.2 an n-dimensional input generates a 3-dimensional output. A commonly used notation for weights is $w_{i,j}^l$, which stands for the weight associated with the i-th neuron at layer l receiving input from the j-th neuron at layer $l-1$. So, for example, $w_{1,1}^2$, $w_{1,2}^2$, and $w_{1,3}^2$ would be the weights for the first neuron (i.e. the top one) at layer 2. A similar notation holds for the biases. Thus b_2^2 would be the bias of the second neuron (i.e., the bottom one) at layer 2. Under supervised learning, a set of inputs with desired targets is given $\{(p_1, t_1), \ldots, (p_n, t_n)\}$. Here $p_1, \ldots, p_n$ denote inputs (thus, p_i is the i-th input vector) and $t_1, \ldots, t_n$ denote the desired targets for these inputs. Initially the weights and biases are set somewhat arbitrarily (usually to small random numbers) and are then continuously adjusted in terms of the error produced. This process is called learning under supervision.

We now outline a small sample of existing neural net structures.

2.1.1 Adaline

The Adaptive Linear Element (or Adaline) is the simplest example of a neural net. It is shown in Figure 2.3.

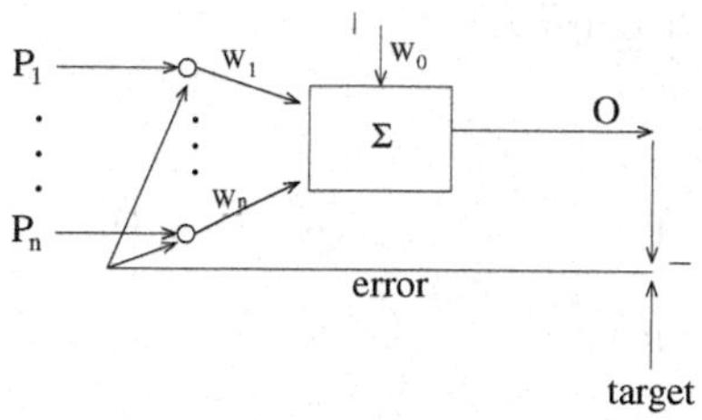

Fig. 2.3

In Figure 2.3 we have one neuron and we have denoted its bias by w_0. The transfer function is taken to be the identity function $O = f(N) = N$. Thus,

$$O = \sum_{i=1}^{n} w_i p_i + w_0.$$

For each input $(p_{1,q}, \dots, p_{n,q})^T$ we specify a target t_q, $q = 1, \dots, Q$. For the q-th input, the squared error is

$$E_q = (t_q - O_q)^2.$$

We would like to adjust the weights and the bias so as to minimize E_q. The partial derivatives are

$$\frac{\partial E_q}{\partial w_i} = -2(t_q - O_q)\frac{\partial O_q}{\partial w_i} = -2(t_q - O_q)p_{i,q}.$$

In order to minimize E_q we take small steps opposite ∇E_q (the gradient of E_q). If $w_1(0), \dots, w_n(0)$, $w_0(0)$ are the initial values given to the weights and the bias, the update from step k to step $k+1$ is given by

$$w_i(k+1) = w_i(k) + \eta(t_q - O_q)p_{i,q}$$

provided input $\mathbf{p}_q = (p_{1,q}, \dots, p_{n,q})^T$ is presented at step $k+1$. Another way to write this is

$$\triangle_q w_i(k) = \eta(t_q - O_q)p_{i,q}, \qquad 0 \le i \le n,$$

where $\triangle_q w_i(k)$ denotes the change of weight w_i at iteration $k+1$ when the q-th input is presented, $t_q - O_q$ is, of course, the error produced when the q-th input is presented, and η is a small positive number.

The Adaline is trained by presenting the input from the training set many times over and updating the weights as shown. Under certain conditions, it can be shown that the weights and the bias will converge to values so that specified inputs will produce outputs close to the specified targets.

2.1.2 Adaptive Nets

Adaptive nets generalize the concept of neural nets in the sense that no weights need to be involved in the computation of the output. Adaptive nets use a training set to update parameters of adaptive nodes. An example of an adaptive net is given in Figure 2.4.

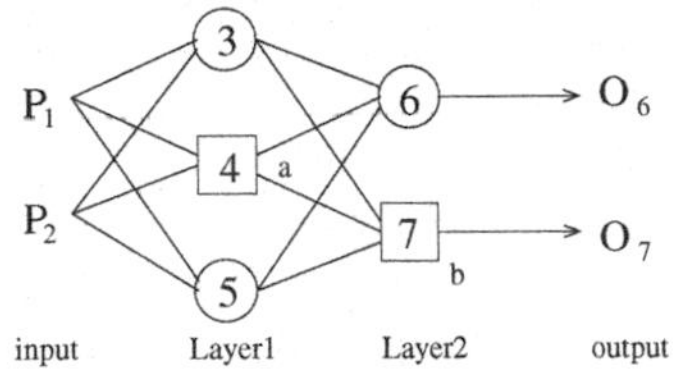

Fig. 2.4

Let $x_{i,j}$ denote the output produced by node j at layer i. The square nodes 4 and 7 produce an output depending on parameters a and b, respectively. The basic idea is to set a and b to some initial values and then use the training set to minimize the sum of squared errors. Let $N(i)$ denote the number of nodes at layer i. In Figure 2.4, $N(0) = 2$, $N(1) = 3$, and $N(2) = 2$. When input $\mathbf{p}$ is presented, the corresponding error function is

$$E_{\mathbf{p}} = \sum_{i=1}^{N(L)} (t_i^{\mathbf{p}} - x_{L,i}^{\mathbf{p}})^2.$$

Here L denotes the last layer, $t_i^{\mathbf{p}}$ denotes the specified target for $\mathbf{p}$ at neuron i of the last layer, and $x_{L,i}^{\mathbf{p}}$ denotes the output of that neuron. To simplify notation we now omit the superscript $\mathbf{p}$. Each sensitivity is defined as

$$\mathcal{E}_{l,i} = \frac{\partial E}{\partial x_{l,i}}.$$

We need to compute the gradient of E, ∇E, and then update the values of the parameters by adding to the current parameter vector $-\eta \nabla E$, where η is a small positive number. Let $f_{i,j}$ be the function giving the output for node $x_{i,j}$. In the example shown in Figure 2.4 we have

$$x_{1,3} = f_{1,3}(p_1, p_2), \quad x_{1,4} = f_{1,4}(a, p_1, p_2), \quad x_{1,5} = f_{1,5}(p_1, p_2),$$
$$x_{2,6} = f_{2,6}(x_{1,3}, x_{1,4}, x_{1,5}), \quad x_{2,7} = f_{2,7}(b, x_{1,3}, x_{1,4}, x_{1,5}),$$
$$O_6 = x_{2,6}, \quad O_7 = x_{2,7}.$$

Furthermore,

$$\nabla E = \left(\frac{\partial E}{\partial a}, \frac{\partial E}{\partial b} \right)^T$$

and

$$\mathcal{E}_{2,i} = -2(t_i - x_{2,i}),$$
$$\mathcal{E}_{1,i} = \frac{\partial E}{\partial x_{1,i}} = \sum_m \frac{\partial E}{\partial x_{2,m}} \frac{\partial x_{2,m}}{\partial x_{1,i}}.$$

So, for our example, we have

$$\mathcal{E}_{2,6} = -2(t_6 - x_{2,6}), \qquad \mathcal{E}_{2,7} = -2(t_7 - x_{2,7}),$$
$$\mathcal{E}_{1,i} = \mathcal{E}_{2,6}\frac{\partial f_{2,6}}{\partial x_{1,i}} + \mathcal{E}_{2,7}\frac{\partial f_{2,7}}{\partial x_{1,i}}, \qquad i = 3, 4, 5,$$
$$\mathcal{E}_{0,i} = \mathcal{E}_{1,3}\frac{\partial f_{1,3}}{\partial p_i} + \mathcal{E}_{1,4}\frac{\partial f_{1,4}}{\partial p_i} + \mathcal{E}_{1,5}\frac{\partial f_{1,5}}{\partial p_i}, \qquad i = 1, 2.$$

Thus,

$$\frac{\partial E}{\partial a} = \frac{\partial E}{\partial x_{1,4}} \frac{\partial f_{1,4}}{\partial a} = \mathcal{E}_{1,4}\frac{\partial f_{1,4}}{\partial a}$$

and similarly

$$\frac{\partial E}{\partial b} = \mathcal{E}_{2,7}\frac{\partial f_{2,7}}{\partial b}.$$

This completes the computation of ∇E and, therefore, the necessary updates of a and b when input p is presented. This process is called backpropagation, as the computation of the sensitivities proceeds from the last layer to the first layer.

2.1.3 The Standard Backpropagation

The standard backpropagation typically refers to a special case of adaptive nets where the parameters are weights and biases and the output produced by neuron i at layer m, $0 \le n \le L$, is given by

$$a_i^m = f_m(n_i^m).$$

Here f_m refers to the transfer function at layer m (applied to each neuron at layer m) and n_i^m is the net input to neuron i at layer m. So,

$$n_i^m = \sum_j w_{i,j}^m a_j^{m-1} + b_i^m.$$

Here a_j^{m-1} refers to the output of neuron j at layer $m-1$, $w_{i,j}^m$ is the weight connecting neuron j at layer $m-1$ to neuron i at layer m, and b_i^m is the bias of neuron i at layer m. The sensitivities $\mathcal{E}_{l,i}$ defined in the previous subsection are now replaced by sensitivities defined as

$$\zeta_i^m = \frac{\partial \hat{E}}{\partial n_i^m},$$

where $\hat{E} = \mathbf{e}^T(k)\mathbf{e}(k)$ and $\mathbf{e}(k)$ is the error vector at layer L (i.e., the output layer) committed on trial k. Using the chain rule in a manner somewhat similar to that in Section 2.1.2, it can be shown that

$$\zeta_i^L = -2(t_i - a_i^L) f_L'(n_i^{L-1}),$$
$$\zeta_i^m = \sum_j w_{j,i}^{m+1} f_m'(n_i^m) \zeta_j^{m+1}.$$

Thus, as in the previous subsection the sensitivities ζ_i^{m+1}, $0 \leq m \leq L-1$, are computed starting from layer L and proceeding step by step to previous layers. The updates are given by

$$w_{i,j}^m(k+1) = w_{i,j}^m(k) - \eta \zeta_i^m a_j^{m-1},$$
$$b_i^m(k+1) = b^m(k) - \eta \zeta_i^m.$$

As earlier, the inputs in the training set are presented over and over with appropriate updates given above. If the positive number η is small enough, the weights and biases will often converge (albeit somewhat slowly) to the specified targets.

Multilayers nets are very flexible. For example, it can be shown that a two-layer network can approximate any reasonable function where the transfer function for the first layer is LogSig and the transfer for the second layer is the identity function. This can be done with any degree of flexibility provided there are enough neurons in the first layer. Even if there are enough neurons, the net may not be able to approximate a given function if the weights and biases are poorly chosen. The reason the net may not converge is that the function to be minimized (i.e., the square of the error) would possibly be highly nonlinear and not quadratic and hence have many local minima. It is good practice to apply the algorithm outlined earlier starting with a number of different initial conditions. It is also important to choose the training set to be representative of the total dataset so that the net is able to successfully generalize the input-output relation to that total dataset. Also, it is worth pointing out that if too much flexibility is given to the net (i.e., too many neurons), the net may generalize in a way not intended by the designer.

Another problem is that if the learning factor η is taken to be too large, then convergence may fail. If η is taken too small, convergence may be too slow for many applications. Some of the methods to address this problem include the use of methods different from the steepest descent (e.g., the conjugate gradient or the Levenberg-Marquard algorithms). Other methods use heuristic techniques and we now mention a few of these. If the learning factor η is taken too large, the weights and biases may oscillate and fail to converge. To reduce the oscillations, one may use a first order filter. The filter works as follows: If at time k the input is $w(k)$, then the output is given by

$$y(k) = \gamma y(k-1) + (1-\gamma)w(k),$$

where $0 \leq \gamma \leq 1$. Note that $\sum y(k) = \gamma \sum y(k-1) + (1-\gamma) \sum w(k)$. Therefore, $E(y) = \gamma E(y) + (1-\gamma)E(w)$, that is

$$E(y) = E(w).$$

Thus, the average value of y is equal to the average value of w. Now, the variation of the weights and biases for the standard backpropagation is

$$\triangle w^m(k) = -\eta \zeta^m (a^{m-1})^T,$$
$$\triangle b^m(k) = -\eta \zeta^m.$$

Thus, at time k the variations are redefined as

$$\triangle w^m(k) = \gamma \triangle w^m(k-1) - (1-\gamma)\eta \zeta^m (a^{m-1})^T,$$
$$\triangle b^m(k) = \gamma \triangle b^m(k-1) - (1-\gamma)\eta \zeta^m.$$

The average values of the oscillations for the weights and the biases remain the same but the amplitude is reduced. In fact, if γ is close to 1, $\triangle w^m$ and $\triangle b^m$ change very little from time $k-1$ to time k.

Another heuristic technique to speed up convergence is to use a variable learning rule. For example, if the squared error (over the entire training set) increases by say more than 5%, then the update is discarded and the learning value η is multiplied by some number less than 1, say by 0.75, and the momentum γ is set to zero. If the squared error decreases after the weight update, then the update is accepted and the learning rule is multiplied by a number greater than 1, say by 1.5, and the momentum γ is reset to its original value. Finally, if the squared error increases by less than 5%, the weight update is accepted but the learning rule and the momentum are left unchanged. The motivation to accept an increase in error (less than 5%) is to be able to explore different values for weights that may potentially lead to a better value although initially the error is slightly higher. Multiplying the learning factor by 1.5 speeds up the updates. On the other hand multiplying the learning rule by 0.75 and setting the momentum to zero is mathematically expressing caution about moving on the current path to an optimal point. It should be pointed out that sometimes this approach fails to obtain convergence even though the standard approach gives convergence. Most of the time, however, this heuristic method yields faster convergence than the standard algorithm.

2.1.4 A Brief Overview of Additional Networks

In this subsection we present a brief overview of a few additional networks.

Counter-Propagation

The Counter-Propagation net is based on joining a Kohonen net with an outstar net. Such a net is sketched in Figure 2.5.

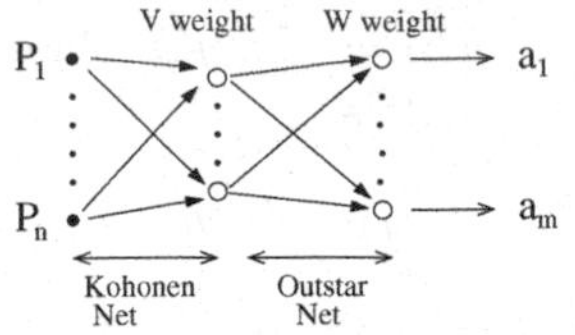

Fig. 2.5

Counter-Propagation sets up a map between n-dimensional vectors and m-dimensional vectors. The neurons at the Kohonen level are competitive. This means that the neuron that has the highest net input is the only neuron that fires and updates its weight. The other neurons have activation zero and do not update their weights. All inputs are first normalized so the highest net input corresponds to the cosine made by the input vector and the neuron's weight vector. Thus, the highest net input corresponds to the best match in direction between the input and the neuron's weight. If neuron i at the Kohonen level is the winner, its weight vector is updated at time k by

$$\mathbf{v}_i(k) = \mathbf{v}_i(k-1) + \eta(\mathbf{p} - \mathbf{v}_i(k-1))$$

if input $\mathbf{p}$ has been presented. The equilibrium position is $\mathbf{v}_i = \mathbf{p}$, or more accurately, $\mathbf{v}_i$ converges to the center of gravity of inputs similar to $\mathbf{p}$ (i.e., inputs for which the same neuron i is the winner). If O_j denotes the output of neuron j at the Kohonen level, then

$$O_j = \begin{cases} 1 \text{ if neuron } j \text{ is the winner} \\ 0 \text{ otherwise.} \end{cases}$$

The second net is the outstar net. The update of the weights at that level is given by

$$\Delta w_{i,j} = \eta'(t_i - w_{i,j})O_j,$$

where t_i represents the i-th component of the desired target. At equilibrium, $w_{i,j} = t_i$ if j is the winning neuron. Since the transfer function is linear (i.e., $f(N) = N$) and $O_j = 0$ except for the winning neuron where $O_j = 1$, the net input to neuron i is $w_{i,j}$. So, $a_i = t_i$ at equilibrium. The right target is obtained at equilibrium. For further information on Counter-Propagation we refer the reader to [9] and [39].

Adaptive Resonance Theory (ART)

We first discuss the outstar nets and instar nets and show how memory patterns might be accessed. The outstar net is shown in Figure 2.6.

As shown, the neuron $\mathbf{p}$ sends an output of 1 to all the neurons in the next layer. The weight vector is made equal to the specified output $(a_1, \ldots, a_n)^T$.

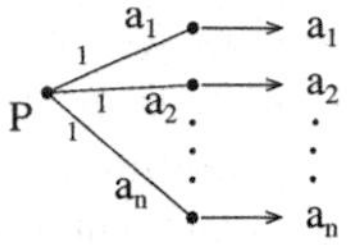

Fig. 2.6

The transfer function here is, of course, linear. Thus, when **p** is active, a specified output vector is emitted by the output layer. The instar net is shown in Figure 2.7.

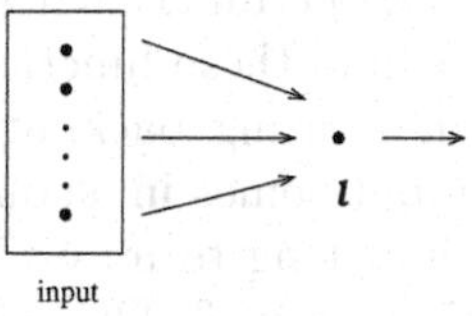

Fig. 2.7

Here, neuron l is only activated by certain inputs. The input **p** needs to be close enough to the weight vector of neuron l for neuron l to be activated. Often instars are structured in a competitive layer, as the following net shows. In order to access memory patterns a competitive layer is paired off with outstar structures, as shown in Figure 2.8.

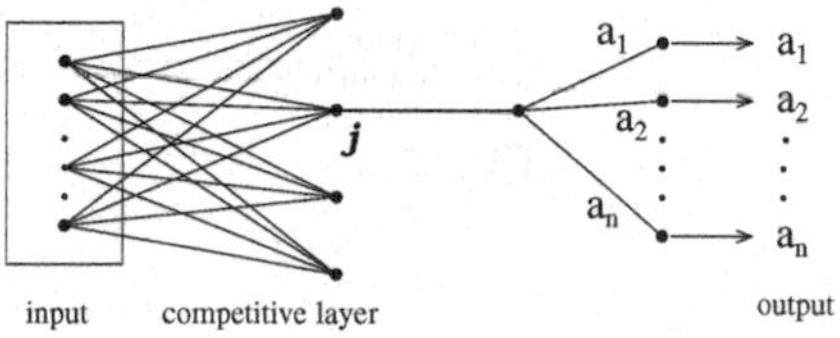

Fig. 2.8

Here, the vector input **p** is sent to the competitive layer. The competition works as follows: The only neuron to be activated is the one where the net input is the largest. The neuron is labeled j in Figure 2.8. Neuron j sends an output of 1 to the next layer, where the weight vector $(a_1, \ldots, a_n)^T$ is chosen to be the vector that one wishes to associate with **p**. This structure is thus able to build an association between vectors. At the end of the process neurons in the competitive layer will represent "clusters" (i.e., will be activated by vectors falling into a similar class). One could slightly change the philosophy of competition by having several winners (the weight of each winner determined by the strength of match between its weight vector and the input vector). This then will allow the user "interpolation" to determine the output.

We now have the background to define the ART structure. Inputs can be binary or analog. For simplicity, we focus the discussion on binary inputs. The analog case is very similar in its broad features but technically more complex. One problem with the classical backpropagation algorithm is that once a net is trained, no incremental training is possible. If an additional (input, output) pair is to be added to the training set at a later time, the network must be retrained on the old training set with the new pair added. Using only the new pair will train the net on that pair only and hence render the net invalid for the initial training set. The ART structure is a complex dynamical structure and, for this reason, we focus on the functional aspects of ART rather than the mathematical machinery behind these functional aspects.

A common problem with nets using backpropagation is the stability or plasticity problem: When an input comes in, should it be classified as similar to previous inputs (i.e., does it belong to an established cluster or should a new cluster be initiated for that input)? The ART structure is sketched in Figure 2.9.

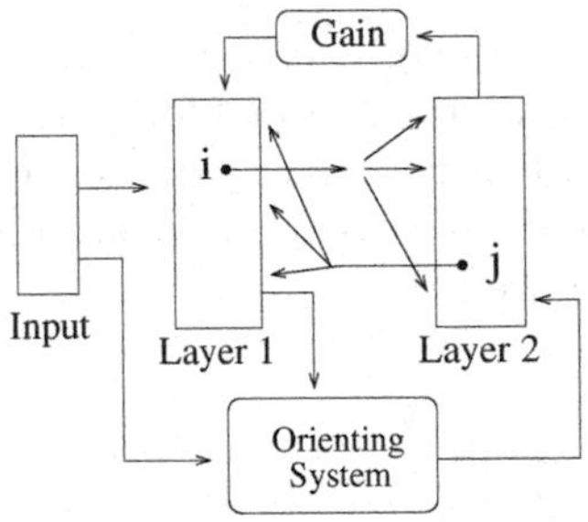

Fig. 2.9

Such a structure has limited capability to generalize from the training set. Because of its capability to perform incremental training, it is useful in spatio-temporal pattern recognition. The main idea is to use "resonance" between Layer 1, the input layer, and Layer 2, the output layer. Layer 1 receives and holds the input pattern. Layer 2 sends a response to Layer 1. If the response is similar to the input, then there is a match. If response does not match them, the two layers will resonate, seeking a match. If the input fails to match any pattern stored in the neurons of Layer 2 within a specified tolerance, a new stored pattern is formed. The orienting system disables the winning neuron (labeled j) in Layer 2 and a new competition takes place without the winner. A new match is attempted with the new winner. If the match does not take place within the described tolerance, the orienting system disables the new winner and a competition takes place without the two previous winners. If a no-point tolerance is met, the input is then viewed as new and an unused neuron in Layer 2 stores its pattern. The connection from neurons in Layer 1 to a neuron of Layer 2 forms an instar net, with Layer 2 being a competitive layer. Competition here has a slightly different meaning, in that normalization of inputs does not take place. Some form of normalization is performed by Layer 1. The connections from a neuron in Layer 2 to neurons in Layer 1 form an outstar net. Denote by V the connections from Layer 1 to Layer 2 and by W the connections from Layer 2 to Layer 1. The match takes place within tolerance α if the following inequality holds:

$$\sum_i w_{j,i} a_i / \sum_i a_i > \alpha.$$

Here, a_i represents the activation of neuron i. The left-hand side of the inequality reflects how similar the (binary) input vector is to the response. If, for example,

$$(a_1, a_2, a_3, a_4) = (1, 0, 1, 1)$$

and

$$(w_{j,1}, w_{j,2}, w_{j,3}, w_{j,4}) = (0, 0, 1, 1)$$

then the left-hand side is 2/3 (i.e., two out of the three 1's of the input match the response). The gain unit outputs 1 if at least one component of the input is 1. Each neuron in Layer 1 has three inputs: the data input, the gain unit, and the response sent by j. For a neuron in Layer 1 to output 1, at least two of the three inputs must be 1 (the two-third rule). If any component in the response is 1, the gain unit is forced to 0. Thus, a neuron in Layer 1 will fire only if its input matches the response (both are 1). If the match test is successful, the input is then associated with the winning neuron in Layer 2. The W weights are updated. The V weights are similarly updated but with normalization

$$w_{i,j}(k+1) = w_{i,j}(k) a_i,$$
$$v_{j,i}(k+1) = \frac{L w_{i,j}(k) a_i}{L - 1 + \sum_k w_{k,j} a_k},$$

where L is some constant. Thus, $v_{j,i}$ is a scaled-down version of $w_{i,j}$. The reason $v_{j,i}$ are updated in a normalized way is to avoid normalizing the prototypes.

For further information on the ART structure, we refer the reader to [5] and [39].

Hebbian Learning

Hebbian learning is inspired by a natural law formulated by Donald Hebb:

> When the axon of a cell A is near enough to excite cell B and repeatedly takes part on firing it, some changes take place in one or both cells such that the efficiency of A firing B is increased. [A need not be the only cell involved in firing B.]

Another way to formulate the above law is as follows: If two neurons on either side of a synapsis are activated simultaneously, the strength of that synapsis increases. Consider the situation sketched in Figure 2.10.

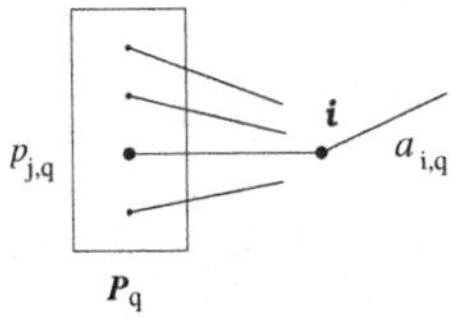

Fig. 2.10

Input $\mathbf{p}_q$ is presented at time $k+1$. Let $p_{j,q}$ denote the j-th component of $\mathbf{p}_q$, $1 \leq q \leq Q$. Let $a_{i,q}$ be the action of neuron i (located at the next layer). Then

$$w_{i,j}(k+1) = w_{i,j}(k) + \alpha a_{i,q} p_{j,q}.$$

Often, α is taken to be 1. Also, if we focus on supervised learning, $a_{i,q}$ is replaced by $t_{i,q}$ (the i-th component of the specified target $\mathbf{t}_q$ for input $\mathbf{p}_q$). The update is then

$$w_{i,j}(k+1) = w_{i,j}(k) + t_{i,q} p_{j,q}.$$

This is conveniently written in matrix form as

$$W = \sum_{q=1}^{Q} \mathbf{t}_q \mathbf{p}_q^T,$$

provided we assume that the initial values of $w_{i,j}$ are all 0 and we have gone through the cycle of presenting inputs $\mathbf{p}_1, \ldots, \mathbf{p}_Q$. Note that if the dimension of the input is r and the dimension of the target is s, then W is an $r \times s$ matrix. This in turn can be conveniently rewritten as

$$W = TP^T,$$

where T denotes a matrix whose columns are the target vectors $\mathbf{t}_1, \ldots, \mathbf{t}_Q$ and P denotes the matrix whose columns are the inputs vectors $\mathbf{p}_1, \ldots, \mathbf{p}_Q$. We consider here a linear associator. This means that if the input is $\mathbf{p}$, the resulting activation is $W_\mathbf{p}$. Assume that the inputs form an orthonormal set; then

$$\mathbf{a}_q = W_{\mathbf{p}_q} = \left(\sum_j \mathbf{t}_j \mathbf{p}_j^T\right) \mathbf{p}_q = \sum_j \mathbf{t}_j (\mathbf{p}_j^T \mathbf{p}_q).$$

Since all the terms $\mathbf{p}_j^T \mathbf{p}_q$ are 0 except when $j = q$, in which case the value is 1, we have

$$\mathbf{a}_q = \mathbf{t}_q.$$

In the case when the inputs form an orthonormal set, the linear associator produces an exact recall, and its activation matches the target. If the inputs are normalized (i.e., $||\mathbf{p}_q|| = 1$) but not orthogonal, then

$$\mathbf{a}_q = \mathbf{t}_q + \sum_{j \neq q} \mathbf{t}_j (\mathbf{p}_j^T \mathbf{p}_q).$$

An error in the recall takes place and that error depends on the amount of correlation between the inputs of the training set.

One way to produce a better recall is to use the pseudo-inverse method. If the number of rows of P is greater than the number of columns of P (i.e., the number of inputs in the training set is less the dimension of the inputs) and the inputs are linearly independent, then one seeks to minimize the squared error $||T - WP||^2$. It can be shown that under the above assumptions the minimum is given by

$$W = TP^+,$$

where P^+ denotes the pseudo-inverse of P, that is, $P^+ = (P^T P)^{-1} P^T$. Note that in the (very unlikely) case when P^{-1} exists then $P^+ = P^{-1}$. Note also that, in contrast to previous structures, no learning takes place. The weights are set to TP^T, or TP^+ if the pseudo-inverse method is required.

Particle Swarm Optimization (PSO)

An interesting methodology that, in particular, could be used to train neural nets was developed by Kennedy and Eberhart in 1995, [4, 16]. For applications using this approach, we refer to [4] and [16]. The PSO approach is similar to genetic algorithms (see a later section) in that the starting point is a population, more or less randomly selected of potential solutions. The difference, in contrast to genetic algorithms, comes in assigning random velocities to each member of the population. Each member of the population is called a particle. Particles are then moved toward a near optimal solution by evaluating the "fitness value" of each particle. In fact, a possible set of equations describing the dynamics of the solution is

$$v_{i,d}(\text{new}) = v_{i,d}(\text{old}) + c_1 \,\text{rand}(\,)(p_{i,d} - x_{i,d}) + c_2 \,\text{rand}(\,)(p_{g,d} - x_{i,d}),$$
$$x_{i,d}(\text{new}) = x_{i,d}(\text{old}) + v_{i,d}.$$

Here, x_i refers to the position of the i-th particle and $x_{i,d}$ refers to the d-th component of x_i. The variables $p_{i,d}$ and $p_{g,d}$ refer to the best position so far occupied by the i-th particle and the global best position over all particles, respectively. The constants c_1 and c_2 have often been set to some values close to 2. The velocities in each direction are bounded by some constant $V_{\max}$. If $V_{\mathbf{max}}$ is high, particles may move past the optimal point. If $V_{\mathbf{max}}$ is too low, exploration becomes limited. The concept of best here refers to relative to some fitness function. The function rand( ) refers to a randomly generated number between 0 and 1. The algorithm then keeps track of the best for each particle and the global best, both of these, of course, updated as needed. The differences $p_{i,d} - x_{i,d}$ and $p_{g,d} - x_{i,d}$ are the distances of the current position of the particle i to its best and to the global best, respectively. The first previous update can be replaced by

$$v_{i,d} = w \times v_{i,d} + c_1 \text{ rand}()(p_{i,d} - x_{i,d}) + c_2\text{rand}()(p_{g,d} - x_{i,d}),$$

where w denotes an inertia coefficient. With that new formulation, $V_{\max}$ can be taken to be the dynamic range of each variable (thus $V_{\max}$ here is a vector). To optimize the weights in a neural net, the variable $x_{i,d}$ consists of all current values for all weights and biases. Additional structure can be incorporated. For example, if the transfer function is $\text{LogSig}(N) = 1/(1 + e^{-cN})$, then the parameter c can be thrown in as one component of the variables $x_{i,d}$. The fitness function could be the sum of squared errors computed over some fixed training set.

2.1.5 Conclusion

We have presented a small sample of some of the important types of neural nets whose weights are determined by a training set. Once a network is trained, given some input that is not part of the training set, the system produces an output. The system is a black box. In general, it is difficult to know what exactly the net has learned. An important property is that neural nets have the capability to adapt. The weights change so that the system gradually learns to reproduce the specified input-output pairs. It is important to choose the right training set so that the net is able to generalize to the whole dataset.

2.2 Fuzzy Logic

Let X denote the universal set. The concept of a "fuzzy subset of X" is a generalization of subset of X. Let A be any subset of X. The subset A can be identified with its characteristic function, which we will often denote by the same symbol A. Thus, the subset A viewed as a characteristic function is defined by

$$A(x) = \begin{cases} 1 & \text{if } x \in A \\ 0 & \text{if } x \notin A. \end{cases}$$

The values taken by A are either 1 or 0. If $A(x) = 1$, then x belongs to A, and if $A(x) = 0$, then x does not belong to A. Thus, A viewed as a characteristic function is "a membership function" since the value of A at x indicates if x is a member of A or not.

We define A to be a fuzzy subset of X if A is a function defined on X that takes values in $[0, 1]$. For example, $A(x) = 0.25$ indicates that the membership of x in A is 0.25 - intuitively x does not belong very much to A. The value $A(x)$ can be interpreted as "on a scale of 0 to 1, how much x fits A." Often, fuzzy sets have a linguistic interpretation. In Figure 2.11 we show an example of how the set "Around 10" might be defined.

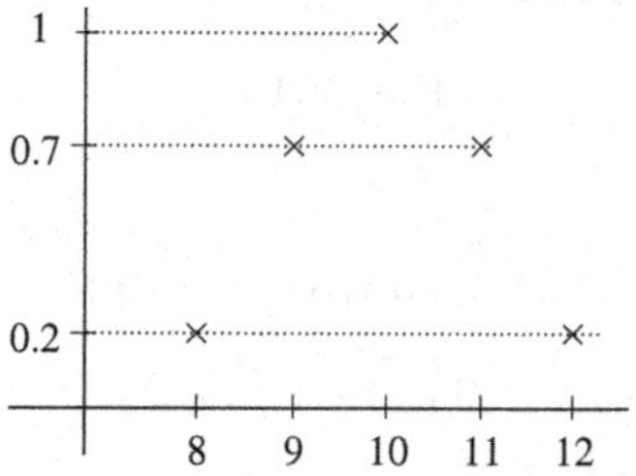

Fig. 2.11

Here, the universal set, often called the "universe of discourse," is $X = \{6, \ldots, 12\}$. The membership of 10 in that set is 1 (i.e., 10 fits perfectly "Around 10"). The memberships of 9 and 11 fit 0.7 the concept of "Around 10." The memberships of $6, 7, 13$, and 14 are 0; these numbers are too far away from 10 to fit the concept of "Around 10." Similarly Figures 2.12, 2.13, and 2.14 could represent "Young," "Old," and "Few," respectively.

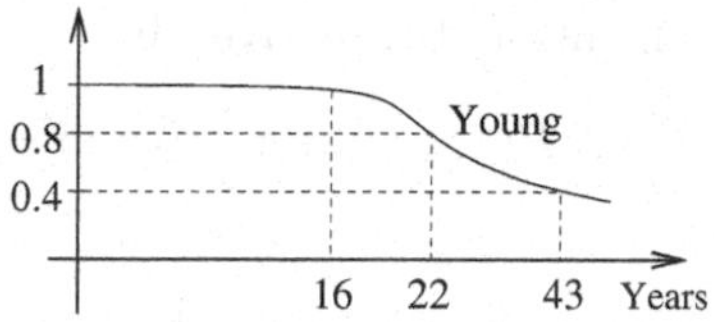

Fig. 2.12

If the universe of discourse X is finite and A is a fuzzy subset of X, the notation

$$A = \sum_i \alpha_i / x_i, \qquad \alpha_i \in [0, 1],$$

signifies that A is the supremum of functions f_i such that $f_i(x_i) = \alpha_i$ and $f_i(x_j) = 0$ if $j \neq i$. In particular, if all the x_i are distinct, $A(x_i) = \alpha_i$. If $x_i = x$ for several distinct values of i, then

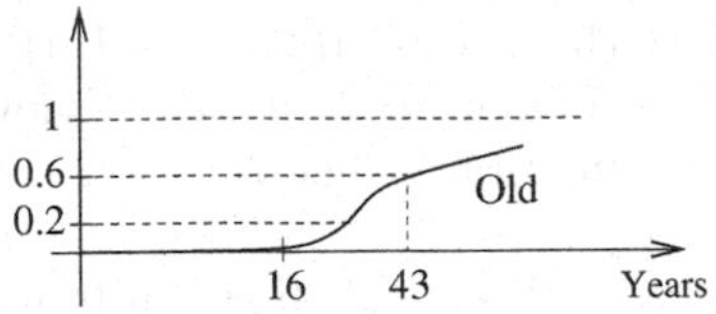

Fig. 2.13

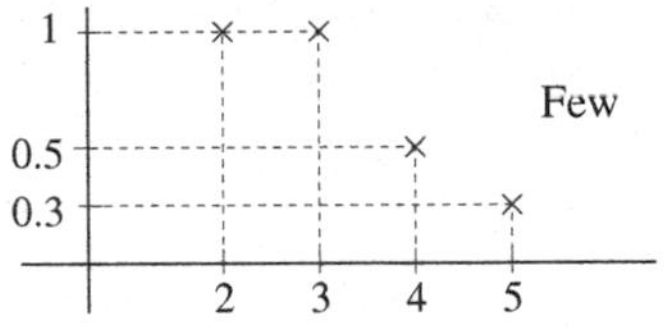

Fig. 2.14

$$A(x) = \sup\{\alpha_i \,|\, x_i = x\}.$$

For example $0.7/x_1 + 0.3/x_1 + 0.5/x_2 + 0.9/x_2 = 0.7/x_1 + 0.9/x_2$. For the continuous case, notation such as

$$A = \int_{x \in [5,10]} e^{-3(x-1)^2} / x$$

signifies that $X = [5, 10]$ and that for each $x \in X$, $A(x) = e^{-3(x-1)^2}$. In the notation $A = \sum_i \alpha_i / x_i$, if $x_j \in X$ and $x_j \neq x_i$ for all i, it is assumed that $A(x_j) = 0$.

2.2.1 Operations of Fuzzy Sets

If A and B denote two subsets of X, then the characteristic functions of $A \cap B$, $A \cup B$, and $\bar{A}$ (the complement of A) are given by

$$(A \cap B)(x) = A(x) \wedge B(x), \qquad (A \cup B)(x) = A(x) \vee B(x),$$
$$\bar{A}(x) = 1 - A(x).$$

Here, $\wedge$ and $\vee$ denote the operations Min and Max, respectively. This definition extends without any change to the case where A and B are fuzzy subsets of X, for example,

$$(0.3/x_1 + 0.8/x_2) \cap (0.4/x_1 + 0.7/x_2) = 0.3/x_1 + 0.7/x_2$$

and

$$(0.3/x_1 + 0.8/x_2) \cup (0.4/x_1 + 0.7/x_2) = 0.4/x_1 + 0.8/x_2.$$

These operations on fuzzy subsets can be generalized by introducing functions from $[0, 1]^2$ into $[0, 1]$ called t-norm and s-norm. For additional information on t-norms and s-norms, see [43].

A t-norm is a function from $[0,1]^2$ into $[0,1]$ satisfying the following:
1. $t(x,y) = t(y,x)$,
2. $t(t(x,y),z) = t(x,t(y,z))$,
3. t is monotone and nondecreasing,
4. $t(0,x) = 0$, $t(1,x) = x$.

Clearly, $\wedge$ satisfies the four properties. An s-norm is a function from $[0,1]^2$ into $[0,1]$ satisfying items 1 to 3 and 5. $s(x,0) = x$, $s(1,x) = 1$. Clearly, $\vee$ satisfies these four properties. Thus, a more general definition of intersection and union of fuzzy sets could be

$$(A \cap B)(x) = t(A(x), B(x)), \qquad (A \cup B)(x) = s(A(x), B(x)),$$

where t and s are arbitrary t-norm and s-norm. There are many examples of t-norms and s-norms, we list a few:

$$t(x,y) = xy, \qquad t(x,y) = 1/(1/x^p + 1/y^p - 1)^{1/p}$$

and

$$s(x,y) = x + y - xy, \qquad s(x,y) = 1 - 1/((1-x)^{-p} + (1-y)^{-p} - 1)^{1/p}.$$

Once a t-norm (or s-norm) is defined, the dual s-norm (or dual t-norm) can be constructed by $s(x,y) = 1 - t(1-x, 1-y)$ (or $t(x,y) = 1 - s(1-x, 1-y)$). If dual norms are used, then DeMorgan's laws hold for fuzzy subsets (i.e., $\overline{(A \cup B)} = \bar{A} \cap \bar{B}$ and $\overline{(A \cap B)} = \bar{A} \cup \bar{B}$). Just as intersections and unions can be generalized using the t and s-norms, the complement can be generalized using the negation operator. A negation operator is a nonincreasing map from $[0,1]$ to $[0,1]$ with the additional properties

$$N(0) = 1, \qquad N(1) = 0.$$

Clearly, $N(x) = 1 - x$ is a special case. Other examples could be

$$N(x) = \frac{1-x}{1+\underline{a}x}, \qquad \underline{a} > -1,$$

or

$$N(x) = (1 + x^w)^{1/w}, \qquad w > 0.$$

2.2.2 Construction of Membership Functions

In applications it is important to be able to define fairly well the membership functions corresponding to linguistic terms. In a given situation what do terms such as Rich, Poor, Middle Age, ..., etc. mean?

Polling Experts

Perhaps the simplest way to define a membership function is to poll experts. For example, suppose we have 10 experts on aging questions such as "Is 30 middle age?" "Is 40 middle age?" "Is 60 middle age?" The answer to be given by the experts should be Yes or No. The number of Yes answers is counted. If all 10 experts classified 40 as middle age, then the membership of 40 in middle age would be 1. If 6 experts classified 60 as middle age, then the membership of 60 in middle age would be 0.6.

Pointwise Comparison

To determine the membership of each element of the universe of discourse in some linguistic set A, one compares all pairs (x_i, x_j) of the universe of discourse and forms the ratios

$$r_{i,j} = A(x_i)/A(x_j).$$

Thus, if $r_{i,j} = 2$, then the expert has determined that x_i fits twice as well the concept of A than x_j does. Clearly, $r_{i,k} = r_{i,j} r_{j,k}$ and $r_{j,i} = 1/r_{i,j}$. Also, $r_{i,i} = 1$. It can easily be checked that if P denotes the matrix whose entries are $r_{i,j}$ and if $1 \leq i, j \leq n$, then

$$P\mathbf{a} = n\mathbf{a}, \qquad \mathbf{a} = (A(x_1), \ldots, A(x_n))^T.$$

If $\mathbf{a}$ is normalized so that $\sum_i A(x_i) = 1$, then

$$A(x_j) = 1/\sum_i p_{i,j}$$

and this determines the membership function of A. For an extensive discussion of this type of methodology, we refer the reader to [37] and [38].

Parametrized Membership Functions

Often the general form of a membership function is known. In this case, one uses data to determine the parameters. The membership function is of the form $A(\alpha, \mathbf{e}, \ldots, \mathbf{x})$, where $\alpha, \mathbf{e}, \ldots$ are parameters and $\mathbf{x}$ is a vector. Then one seeks to minimize

$$\sum_{i=1}^{n} (A(\alpha, \mathbf{e}, \ldots, \mathbf{x}_i) - y_i)^2$$

as a function of the parameters $\alpha, \mathbf{e}, \ldots$. If one specifies that $A(\alpha, \mathbf{e}, \ldots, x_i) = y_i$ for all i, then, of course, in most cases an analytical solution cannot be obtained and methods such as the steepest descent are used to get an approximate solution. In order to avoid landing on a local minimum, different initial values for $\alpha, \mathbf{e}, \ldots$ are set.

Using Neural Nets

If the value of the membership function is specified at certain points, then this defines a training set and, for example, backpropagation as described earlier could be used. Given a value x in the universe of discourse, the output provided by the net will be interpreted as $A(x)$.

We now list some of the standard membership functions. The triangular function is defined as

$$A(x) = \begin{cases} 0 & \text{if } x \leq a \\ (x-a)/(m-a) & \text{if } a \leq x \leq m \\ (b-x)/(b-m) & \text{if } m \leq x \leq b \\ 0 & \text{if } x \geq b. \end{cases}$$

The triangular function conveys the idea of "around m."

The S-function is defined as

$$A(x) = \begin{cases} 0 & \text{if } x \leq a \\ 2[(x-a)/(b-a)]^2 & \text{if } a \leq x \leq (a+b)/2 \\ 1-2[(x-b)/(b-a)]^2 & \text{if } (a+b)/2 \leq x \leq b \\ 1 & \text{if } x > b. \end{cases}$$

The S-function is shown in Figure 2.15 and it is used for concepts such as Old, Rich, Heavy, Large, ..., etc. The point at which $A(x) = 0.5$ is $(a+b)/2$.

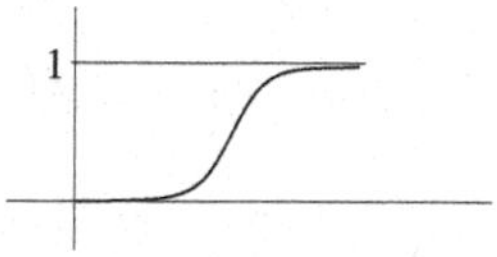

Fig. 2.15

The Gaussian function is defined as

$$A(x) = e^{-c(x-m)^2} \qquad \text{with } c > 0.$$

2.2.3 Fuzzy Relations

Fuzzy relations play a key role in fuzzy inference schemes. The extension from the standard case is very straightforward.

If X and Y denote two universes of discourse, a relation R from X to Y is simply a function from $X \times Y$ into $[0,1]$. Thus, R is simply a fuzzy subset of $X \times Y$. If X and Y are finite, then R can be represented by a matrix whose entries are $R(x_i, y_j)$. Crucial for establishing the fuzzy inference procedure is the sup-min composition. If R_1 is a fuzzy relation from X into Y and R_2 is a fuzzy relation from Y into Z, then we define

$$R_1 \circ R_2(x, z) = \sup_{y \in Y} R_1(x, y) \wedge R_2(y, z).$$

Thus, $R_1 \circ R_2$ is a relation from X into Z, and if R_1 and R_2 are represented by matrices, then the matrix of $R_2 \circ R_1$ is obtained by "multiplying" the matrix of R_1 by the matrix of R_2 where the product is replaced by $\wedge$ and the sum is replaced by taking the supremum. For example,

$$\begin{bmatrix} 0.9 & 0.1 \\ 0.6 & 0.2 \\ 0.4 & 0.5 \end{bmatrix} \circ \begin{bmatrix} 0.7 & 0.4 \\ 0.2 & 0.8 \end{bmatrix} = \begin{bmatrix} 0.7 & 0.4 \\ 0.6 & 0.4 \\ 0.4 & 0.5 \end{bmatrix}.$$

Similarly, if A is a fuzzy set subset of X and R is a fuzzy relation from X to Y, then $A \circ R$ is defined by

$$A \circ R(y) = \sup_{x \in X} A(x) \wedge R(x, y).$$

Of course, $\wedge$ and sup could be replaced by any t-norm and any s-norm, respectively, to obtain a more general operation.

We now proceed to define the implication relation. In general, if A is a fuzzy subset of X and B is a fuzzy subset of Y, $A \to B$ should be defined as a relation from X into Y:

$$(A \to B)(x, y) = f(A(x), B(y)),$$

where f is a function satisfying certain conditions that define an "implication function." Some of the common choices for f are

$$f(x, y) = x \wedge y, \qquad f(x, y) = xy, \qquad f(x, y) = (1 - x) \vee y,$$

and

$$f(x, y) = \sup\{c \in [0, 1] \,|\, x \wedge c \leq y\}, \qquad x, y \in [0, 1].$$

It is easy to see that if we take A and B to be standard subsets of X and Y, then the first two examples of implication functions imply

$$(A \to B)(x, y) = 1 \text{ if and only if } x \in A \text{ and } y \in B.$$

Thus, in the first two cases, $(A \to B) = A \times B$. In the third example, it is easy to see that $(A \to B) = \bar{A} \vee B$ (which is the standard definition of implication) and in the last example. In the third example where $f(x, y) = (1 - x) \vee y$, if one takes $X = Y$ and if A and B are crisp subsets of X, then $(A \to B)$ becomes identical to $A \subset B$. For work dealing with fuzzy relations, see [34] and [1].

2.2.4 Fuzzy Reasoning

If we know that the variable U is some fuzzy subset A' of X and if we know that the rule: If U is A then V is B, where A and B are fuzzy subsets of X

and Y holds, then what inference can be made about V? The answer is to use the sup-min composition:

$$V \text{ is } A' \circ (A \to B).$$

To have an intuitive idea why this works, let A and B be standard sets and let $A' = A$. Then the above formula becomes

$$V(y) = \sup_x A(x) \wedge A(x) \wedge B(y) = B(y).$$

So we obtain the standard implication: U is A and (if U is A then V is B) implies V is B. In particular, if U is a number a, then V is

$$V(y) = \sup_x X_a(x) \wedge A(x) \wedge B(y) = A(a) \wedge B(y).$$

Here, X_a denotes a function defined on X that is 1 if $x = a$ and 0 otherwise.

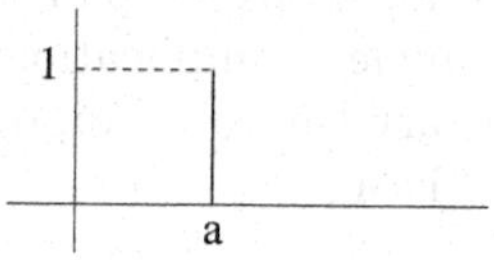

Fig. 2.16a

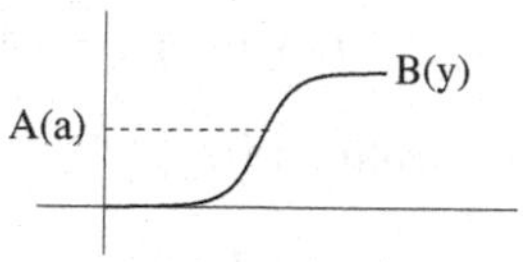

Fig. 2.16b

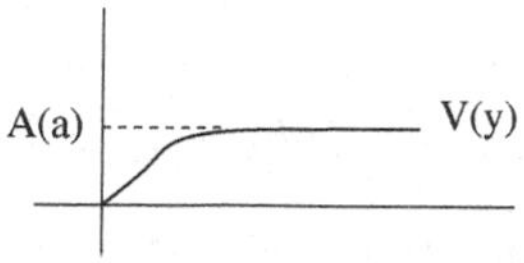

Fig. 2.16c

Figure 2.16a shows U identified with X_a. Figure 2.16b shows the membership function of B. Fig. 2.16.c shows the membership of V when U is a and the presence of the rule: if U is A then V is B.

Similar conclusions are drawn with different choices of the implication function. In the above example, $A(a)$ denotes the strength of the rule given the input $U = a$. If the input is $U = A'$, where A' is a fuzzy subset of X, then

$$V(y) = \sup_x A'(x) \wedge A(x) \wedge B(y).$$

The quantity $\sup_x A'(x) \wedge A(x)$ is called the possibility of A' and A and is denoted by $\text{Poss}(A, A')$. So, $V(y) = \text{Poss}(A, A') \wedge B(y)$. In this case, the strength of the rule (relative to the input U is A') is $\text{Poss}(A, A')$. It should be noted that $\text{Poss}(A, A')$ represents "the largest intersection of A and A'." Of course, the rule may have multiple antecedents and consequents such as:

If U_1 is A_1 and U_2 is A_2 and U_3 is A_3, then V_1 is B_1 and V_2 is B_2.

The input for this rule is U_1 is A'_1, U_2 is A'_2, U_3 is A'_3. The inferred fuzzy set is

$$W(y_1, y_2) = \text{Poss}(A_1, A'_1) \wedge \text{Poss}(A_2, A'_2) \wedge \text{Poss}(A_3, A'_3) \wedge B_1(y_1) \wedge B_2(y_2).$$

It specifies the membership of an arbitrary pair (y_1, y_2) in the fuzzy output W. In the more general case, any legal implication function is used and intersection is replaced by an arbitrary t-norm. Suppose that we have the input

U_1 is A'_1, U_2 is $A'_2, \ldots,$ U_n is A'_n

and the rule

If U_1 is A_1 and U_2 is A_2 ...and U_n is A_n, then V_1 is B_1 and V_2 is B_2 ... and V_m is B_m.

The multiplicative input is defined by

$$P_i(x_1, \ldots, x_n) = t[A'_1(x_1), \ldots, A'_n(x_n)],$$

the multiplicative antecedent is defined by

$$P_a(x_1, \ldots, x_n) = t[A_1(x_1), \ldots, A_n(x_n)],$$

the multiplicative consequent is defined by

$$P_c(y_1, \ldots, y_m) = t[B_1(y_1), \ldots, B_m(y_m)],$$

and

$$R(x_1, \ldots, x_n; y_1, \ldots, y_m) = f(P_a(x_1, \ldots, x_n), P_c(y_1, \ldots, y_m))$$

defines the fuzzy relation generated by the rule. The output under these conditions is then $W = P_i \circ R$; that is,

$$W(y_1, \ldots, y_m) = \sup_{x_1, \ldots, x_n} P_i(x_1, \ldots, x_n) t R(x_1, \ldots, x_n; y_1, \ldots, y_m).$$

In summary, given an if-then rule with fuzzy antecedent and fuzzy consequent, this rule defines a fuzzy relation R. Given then a fuzzy input, the rule produces a fuzzy output obtained by applying the sup-min composition to the (input, rule) pair. Consider the following example:

Rule: If the Pressure is High, then the Volume is Small.

Input: Pressure is Average.

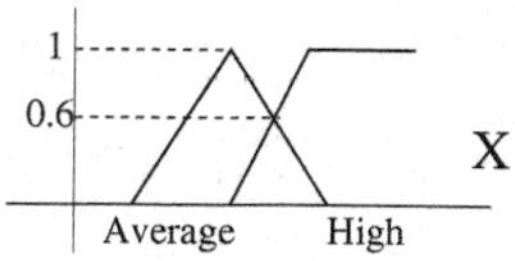

Fig. 2.17

Figure 2.17 shows the memberships of the fuzzy sets High and Average. In this example, $\text{Poss}(\text{Average}, \text{High}) = 0.6$ since this is the largest intersection of Average and High. Figure 2.18 shows the membership function of the fuzzy set $(W(y) = 0.6) \wedge \text{Small}(y)$ if the implication function is $f(x, y) = x \wedge y$ or $(W(y) = 0.6)\text{Small}(y)$ if the implication function is $f(x, y) = xy$.

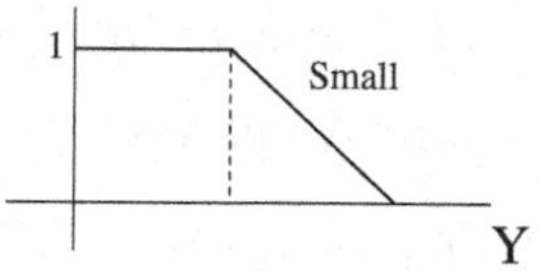

Fig. 2.18

Note that any $y \in Y$ will belong less to the output W than it belongs to Small. Intuitively, Small is implied by High is the only rule we have and the pressure does not match High. This example points to two items that need to be addressed:

1. The output produced is fuzzy. In many situations one needs a single number as an output (e.g., How many cm^3 does the Volume occupy?). This will be addressed in a later section; see defuzzification.

2. It is clear that more than one rule is needed since any single rule is only partially applicable to a given input.

We now assume that we have a set of rules of the form

R_k: If U is A_k, then V is B_k, $\qquad k = 1, \ldots, N.$

The fuzzy input is U is A'. The sets $A_1, \ldots, A_N, B_1, \ldots, B_N$ are in general fuzzy. Each rule produces a fuzzy output as described above. Thus, we have $W_1, \ldots, W_N$, the outputs produced by the N rules. The answer is obtained by aggregating these N outputs:

$$W(y) = \mathcal{A}_{k=1}^{N} W_k(y),$$

where $\mathcal{A}$ denotes an aggregation operation. By an aggregation operation we mean a map $\mathcal{A}$ from $[0,1]^p$ into $[0,1]$, with p is a positive integer - that is, nondecreasing in each of its p variables and

$$\mathcal{A}(0,\ldots,0) = 0, \qquad \mathcal{A}(1,\ldots,1) = 1.$$

An often used aggregation operation is the V operation. Thus, we may define

$$W(y) = V_{k=1}^{N} W_k(y).$$

This may be generalized to multiple antecedents and consequents. Using notations previously defined in this section and generalizing intersection and the V operations to t-norms and s-norms, respectively, we have

$$\begin{aligned}
&P_i(x_1,\ldots,x_n) = t[A_1'(x_1),\ldots,A_n'(x_n)],\\
&P_{a,k}(x_1,\ldots,x_n) = t[A_{1,k}(x_1),\ldots,A_{n,k}(x_n)],\\
&P_{c,k}(y_1,\ldots,y_m) = t[B_{1,k}(y_1),\ldots,B_{m,k}(y_m)],\\
&R_k(x_1,\ldots,x_n;y_1,\ldots,y_m) = f[P_{a,k}(x_1,\ldots,x_n),P_{c,k}(y_1,\ldots,y_m)],
\end{aligned}$$

and then

$$W = P_i \circ R, \quad \text{where} \quad R = \mathcal{A}_{k=1}^{N}.$$

An alternative way to combine the N rules is

$$W_k = P_i \circ R_k \quad \text{and} \quad W = \mathcal{A}_{k=1}^{N} W_k.$$

Thus, the two ways to aggregate are

$$P_i \circ \mathcal{A}_{k=1}^{N} R_k \quad \text{and} \quad \mathcal{A}_{k=1}^{N}(P_i \circ R_k).$$

In general, these two ways generate different outputs. The outputs will coincide if $\vee$ is taken for aggregation and $\wedge$ is taken for t-norm. Consider the following two rules:

If Pressure is High then Volume is Small,

If Pressure is Low then Volume is Large,

and say the input is Pressure is Average. Assume Poss(High, Average) $=$ 0.6 and Poss(Low, Average) $= 0.3$. Then $W_1(y) = 0.6\text{Small}(y)$, $W_2(y) = 0.3\text{Large}(y)$, where the implication function was chosen to be the product, and $R_1(x,y) = \text{High}(x)\text{Small}(y)$, $R_2(x,y) = \text{Low}(x)\text{Large}(y)$, and $R(x,y) = R_1(x,y) \vee R_2(x,y)$.

Much work has been published on "approximate reasoning", examples of which were sketched above. For further reading on approximate reasoning and related matters see [27], [26], [44], [21], [22], [34], and [36].

2.2.5 Other Rules

Although the standard if-then rules are the most commonly used, at times different types of rules are needed. In this section we list a small sample of alternate rules. The first rule we list addresses the problem of the rule working most of the time except when some condition occurs. The general form is

If U is A, then V is B unless W is C.

The input to this rule should be of the form (U, V) is D, where D is a fuzzy subset of $Y \times Z$, Y is the universe of discourse for U, and Z is the universe of discourse for W. The above rule can then be replaced by the following two rules:

(i) If U is A and W is not C, then V is B.

(ii) If U is A and W is C, then V is not B.

Rule (i) is the default case. If W is not C, the rule must hold; that is, if U is A then V is B holds. Rule (ii) expresses the exception i.e. if U is A then we do not want V to be B because W is C. Each of these rules generates a fuzzy relation. That relation will depend on the implication function chosen. Suppose we pick

$$f(x, y) = \begin{cases} 1 \text{ if } x \leq y \\ y \text{ otherwise.} \end{cases}$$

Then the relations generated by rules (i) and (ii) are

$$R_1(x, y, z) = \begin{cases} 1 & \text{if } A(x) \wedge \bar{C}(z) \leq B(y) \\ B(y) & \text{otherwise} \end{cases}$$

and

$$R_2(x, y, z) = \begin{cases} 1 & \text{if } A(x) \wedge C(z) \leq \bar{B}(y) \\ \bar{B}(y) & \text{otherwise.} \end{cases}$$

The relation generated by the two rules is

$$R(x, y, z) = R_1(x, y, z) \wedge R_2(x, y, z).$$

The output generated by the unless rule is

$$V(y) = \sup_{x,z} D(x, z) \wedge R(x, y, z).$$

In particular, if the input is numerical, that is,

$$D(x, z) = \begin{cases} 1 \text{ if } x = x^* \text{ and } z = z^* \\ 0 \text{ otherwise,} \end{cases}$$

it is easy to see that

$$R(x,y,z)=\begin{cases} B(y) \text{ if } B(y) < A(x^*) \wedge \bar{C}(z^*) \text{ and } A(x^*) \wedge C(z^*) \leq \bar{B}(y) \\ \bar{B}(y) \text{ if } A(x^*) \wedge \bar{C}(z^*) \leq B(y) \text{ and } A(x^*) \wedge C(z^*) > \bar{B}(y). \end{cases}$$

Otherwise, $R(x,y,z) = 1$. Intuitively, this expresses the idea that V under different conditions behaves as B or $\bar{B}$.

Another type of rule is a rule that expresses uncertainty about the rule itself. One form of this is

If U is A, then V is B with certainty $\underline{a}$.

Here, $\underline{a} \in [0,1]$. If $\underline{a} = 1$, we are certain about the rule. If $\underline{a} = 0$, we are totally uncertain about the rule. The fuzzy relation generated by this rule is

$$R^*(x,y) = [R(x,y) \wedge \underline{a}] + (1 - \underline{a}).$$

Here, R denotes the relation generated by the rule: If U is A, then V is B. So, $R(x,y) = f[A(x), B(y)]$, where f is the selected implication function. Note that if $\underline{a} = 1$, $R^*(x,y) = R(x,y)$. If, on the other hand, $\underline{a} = 0$, then $R^*(x,y) = 1$. In that case, $V(y) = \sup_x A'(x) \wedge R^*(x,y) = 1$ if we assume that the fuzzy input A' is normal (i.e., $\sup_x A'(x) = 1$). This implies $V = Y$, which reflects total uncertainty, as we are not able to pin down what elements or Y are more or less believed to be in the output V. Intuitively, every element of Y is equally possible as an output in a case of total uncertainty.

Another type of uncertainty rule is one that is truth-qualified. The general form is

If U is A, then V is B is S.

Here, S stands for a fuzzy set representing a truth qualification. An example is If U is A, then V is B is very true. In this context, true is a fuzzy subset of $[0,1]$, whose membership is true$(x) = x$. Very true might have a membership function like Verytrue$(x) = x^2$, and Somewhat true might have a membership function like Somewhattrue$(x) = \sqrt{x}$. The statement If U is A, then V is B is S generates a fuzzy relation R^* defined by

$$R^*(x,y) = S(R(x,y)),$$

where R is the relation defined by the statement If U is A, then V is B. Note that if S = true, then $S(R(x,y)) = R(x,y)$, so $R^* = R$ and the two statements: If U is A then V is B is true and If U is A, then V is B are equivalent statements.

The next type of statement involving uncertainty that we discuss in this section has the form

If Probability(U is A) is P then V is B.

Here, P denotes a fuzzy probability. For example, if P = Likely, P is a fuzzy subset of $[0,1]$ whose membership function might be as shown in Figure 2.19.

Any probability exceeding 0.8 is a perfect example of Likely. The input to such a rule is a probability distribution on X, the universe of discourse of U. Let f be such a probability distribution. Let

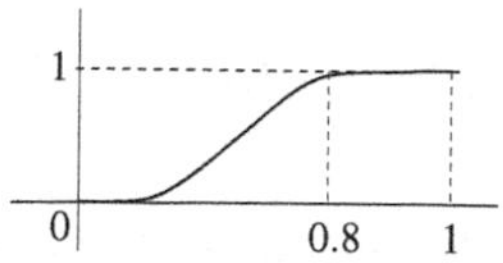

Fig. 2.19

$$W_f = \sum_x A(x) f(x).$$

The quantity W_f represents the average membership in A relative to f. The output is then

$$V(y) = P(W_f) \wedge B(y).$$

What this says is that if the average membership of A relative to the input probability distribution f perfectly fits P, then the output is B.

The last type of statement involving uncertainty we illustrate in this section is one that uses qualifiers. An example of such a rule could be

If Most Big Trucks are Heavy, then the Supply of Gas would be Small.

Here the qualifier is Most. The membership function of Most could be as shown in Figure 2.19. The output in this case could be

$$\text{GasSupply}(y) = \text{Most}\left(\frac{|\text{BigTruckandHeavy}|}{|\text{BigTruck}|}\right) \wedge \text{Small}(y).$$

Here, $|A|$ refers to the cardinality of the fuzzy set A (i.e., $|A| = \sum_x A(x)$). The universe of discourse is the set of trucks under consideration. Each truck has a membership value in Big Truck and in Heavy. The ratio in the parentheses is the fuzzy version of the fraction of Big Trucks that are Heavy.

2.2.6 Defuzzification

In previous subsections we have shown how a set of rules, when provided with some input, generate and output. The output was a fuzzy subset of the appropriate universe of discourse. Most of the time we need to defuzzify the output (i.e., to select an element that in some sense best represents the fuzzy output) Several methods are available and we briefly indicate some of these. In this subsection, B_k will denote the fuzzy output generated by the k-th rule, $1 \leq k \leq N$. We assume that the input is a numerical vector. In a previous subsection we have pointed out that one often aggregates the sets B_k. Let B denote that aggregation. (For example, $B = V_{k=1}^N$ if V is used for aggregation.)

The Centroid Method

The domain of B is partitioned into points $y_1, \ldots, y_M$ and we set

$$y_c(x) = \sum_{i=1}^{M} y_i B(y_i) / \sum_{i=1}^{M} B(y_i).$$

Here, x denotes the input, to stress that the expression on the right-hand side depends on the input x since each B_k and hence B depends on the input x. In practice, the computation of y_i may take some time.

The Center of Sums Defuzzifier

Here, the sets B_k are first combined by addition; that is,

$$B(y) = \sum_{k=1}^{N} B_k(y).$$

Then the centroid of B is found by the formula

$$y_a(x) = \sum_{k=1}^{N} c_{B_k} a_{B_k} / \sum_{k=1}^{N} a_{B_k}.$$

Here, c_{B_k} denotes the centroid for B_k and a_{B_k} denotes the area under B_k. It should be noted in this approach that the areas of overlap between distinct B_k are counted at least twice. However, y_a is easier to compute than y_c.

The Height Defuzzifier

The formula giving this defuzzification is

$$y_h(x) = \sum_{k=1}^{N} \bar{y}_k B_k(\bar{y}_k) / \sum_{k=1}^{N} B_k(\bar{y}_k).$$

Here, $\bar{y}_k$ is the point at which B_k assumes its largest value. If there is more than one point at which B_k is maximum, then $\bar{y}_k$ is the average of all these points.

The Center of Sets Defuzzifier

The defuzzification formula is

$$y_{\mathrm{CoS}}(x) = \sum_{k=1}^{N} c_k \prod_{j=1}^{p} F_j^k(x_j) / \sum_{k=1}^{N} \prod_{j=1}^{p} F_j^k(x_j).$$

Here, it is assumed that the antecedent of the k-th rule is

If X_1 is F_1^k and X_2 is F_2^k and ... and X_p is F_p^k, and the strength of the rule under input $\mathbf{x} = (x_1, \ldots, x_p)^T$ is $\prod_{j=1}^{p} F_j^k(x_j)$. The number c_k denotes the centroid of B_k. Note that if the consequents of all rules are fuzzy sets G_k and each G_k is symmetric, normal, and convex, then the computations are greatly simplified since $c_k = \bar{y}_k$ and $G_k(\bar{y}_k) = 1$. The quantities $\bar{y}_k$ are known ahead of time and we then have $y_{\text{CoS}} = y_h$ for every input x.

For additional information on defuzzification we refer the reader to [10], [25], and [45].

2.2.7 Two Applications

Numerous applications have been studied using fuzzy rules. In this subsection we sketch two rather different applications.

Forecasting Time Series

Given p measurements, we would like to predict what the next one will be. Assume we have N measurements recorded, $x_1, \ldots, x_N$. This generates $N - p$ elements of a training set of the form

$$(x_k, \ldots, x_{p+k-1});\ x_{p+k}, \qquad k = 1, \ldots, N - p.$$

Then we define $N - p$ rules, each rule being defined by one of the training elements

R_k: If X_1 is A_k^1 and ... and X_p is A_k^p, then Y is B_k,

$1 \leq k \leq N - p$. Each A_k^i is a fuzzy set, say a triangular or a Gaussian membership function centered at x_{k+i-1}. The set B_k is centered at x_{p+k}. Using an additional training set together with a least squares fit, one could later refine the parameters of the membership functions.

An alternate approach to the same problem is to locate an interval $[x_L, x_U]$ where all x_i fall. Partition this interval into overlapping subintervals A_1, A_2, ... as shown in Figure 2.20.

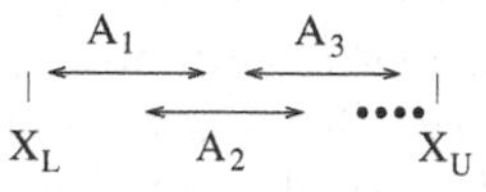

Fig. 2.20

Consider, for example, triangular membership functions, the bases of the triangles being A_1, A_2, ... Determine the degree to which each x_i belongs to the triangular functions. Let the degree be the largest at the triangle based on $A_{j(i)}$. Then the rules generated are

R_k: If X_1 is $A_{j(k)}$ and X_2 is $A_{j(k+1)}$ and ...and X_p is $A_{j(k+p-1)}$, then Y is $A_j(k+p)$,

$1 \leq k \leq N-p$. Conflicts may be present. The same antecedent might be generated by $(x_k, \ldots, x_{k+p-1})$ and by $(x_l, \ldots, x_{l+p-1})$, yet a different consequent generated by x_{k+p} and x_{l+p}. In that case, the conflict might be resolved by selecting the strongest rule (i.e., by comparing $A_{j(k)}(x_k) \wedge \ldots \wedge A_{j(k+p)}(x_{k+p})$ and $A_{j(l)}(x_l) \wedge \cdots \wedge A_{j(l+p)}(x_{l+p})$). Thus, when a sequence such as $y_1, \ldots, y_p$ is fed to the rules, a fuzzy output in produced and then defuzzified as described in the previous subsections. This output will be the predicted next element of the sequence.

A Fuzzy Controller

We have a mobile robot. The goal is to have the robot move so that it avoids obstacles. In this example, the robot has four sensors. The sensors measure distances front, rear, left, and right. The diagram in Figure 2.21 presents the robot under discussion.

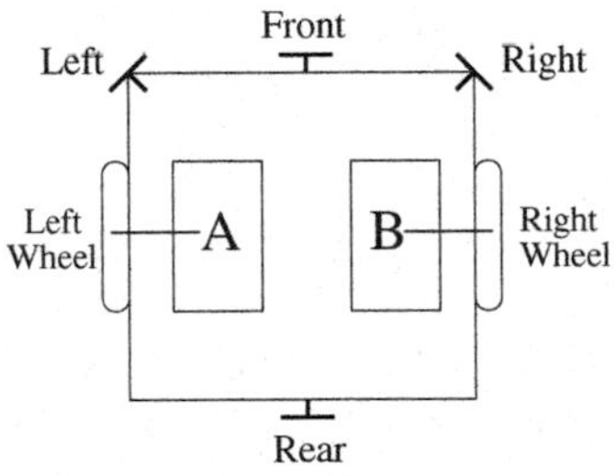

Fig. 2.21

The blocks A and B represent motors contributing to the left and the right wheel. Front, Rear, Left, and Right represent the four sensors that evaluate their respective distances to the obstacle. The distances can have three fuzzy values: Short, Average, and Large. Each motor speed could have seven fuzzy values: Back Fast, Back Average, Back Slow, Stand Still, Forward Slow, Forward Average, and Forward Fast. Rules could be defined as follows:

R_1: If Left Distance is Large and Front Distance is Large and Right Distance is Large, then Left Motor is Fast Forward and Right Motor is Fast Forward.

R_2: If Front Distance is Short, then Left Motor is Forward Average and Right Motor is Backward Average.

The second rule states that if there is an obstacle a short distance ahead, then make a left turn. Other rules such as go to the right if there is an obstacle on the left can easily be written down.

Fuzzy controllers form the bulk of the applications in the area of fuzzy logic. A small fraction of the literature dealing with Fuzzy Controllers can be found in [42], [2], [18], [27], and [26].

2.2.8 The Takayi-Sugeno-Kang Model (TSK Model)

A typical rule in this system is of the form

If X is A and Y is B, then $Z = f(x, y)$.

The input is typically numerical and often the values of the input components are precisely x and y. Often studied are the cases where f is a constant function or a linear function $px + qy + r$. Suppose we have two such rules:

R_1: If X is A_1 and Y is B_1 then Z is $p_1x + q_1y + r_1$.

R_2: If X is A_2 and Y is B_2, then Z is $p_2x + q_2y + r_2$.

If the input is $(x, y)^T$, then the output is

$$\frac{w_1(p_1x + q_1y + r_1) + w_2(p_2x + q_2y + r_2)}{w_1 + w_2},$$

where $w_1 = A_1(x)tB_1(y)$ and $w_2 = A_2(x)tB_2(y)$ and t denotes a t-norm, say $\wedge$ or the product. The interesting part is that no defuzzification is performed; thus the output is not computationally intensive. The TSK model is well adapted to be the fuzzy component of a neuro-fuzzy system. These systems will be discussed in a later section.

2.2.9 Fuzzy Sets of Type 2

A fuzzy set of type 2 is defined by its membership function that maps $X \times [0, 1]$ into $[0, 1]$. Thus, if $\hat{A}$ denotes a fuzzy set of type 2, $\hat{A}(x, u)$ is a number in $[0, 1]$ and the interpretation is: The belief that the membership of x in $\hat{A}$ is u is $\hat{A}(x, u)$. The diagram in Figure 2.22a describes the situation.

The diagram in Figure 2.22b shows a section of $\hat{A}$ when x is kept constant. For every x fixed, $\hat{A}(x, u) = \hat{A}_x(u)$ and $\hat{A}_x$ is a fuzzy subset of $[0, 1]$. In Figure. 2.22b, $\hat{A}_x$ has a Gaussian-like membership function.

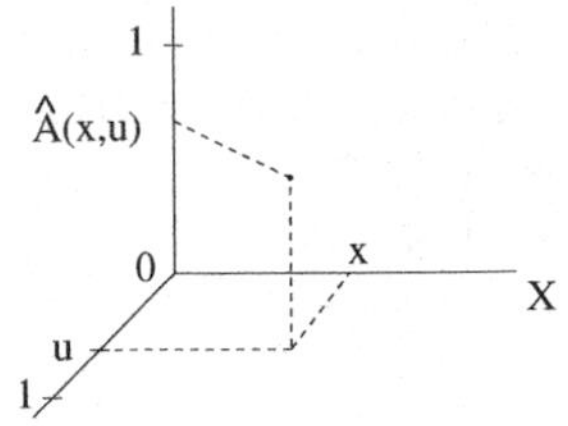

Fig. 2.22a

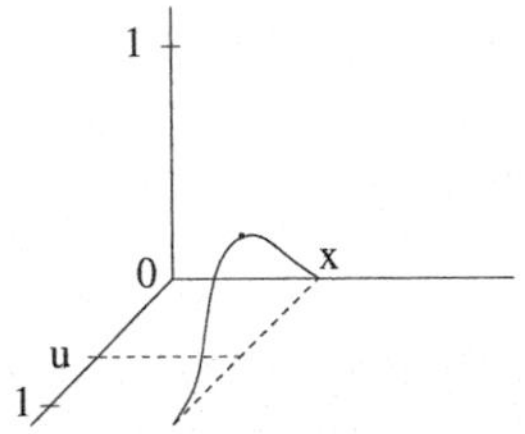

Fig. 2.22b

Thus, $\hat{A}$ factors in the uncertainty that the membership of x in $\hat{A}$ is u, that is, the possibility that x has membership u in $\hat{A}$ is $\hat{A}(x, u)$. Most applications using fuzzy sets of type 2 deal with the case where $\hat{A}_x$ is a subinterval I_x of $[0, 1]$ depending on x; that is

$$\hat{A}_x(u) = \begin{cases} 1 \text{ if } u \in I_x \\ 0 \text{ otherwise.} \end{cases}$$

The diagram in Figure 2.23 depicts a situation where $X = \{1, 2, 3\}$ and a type 2 fuzzy subset $\hat{A}$ of X, where the membership of 1 in X is some number between 0.3 and 0.6 (but it is unknown which number that is), the membership of 2 in $\hat{A}$ is between 0 and 0.3, and the membership of 3 in $\hat{A}$ is 0.6 and 1.

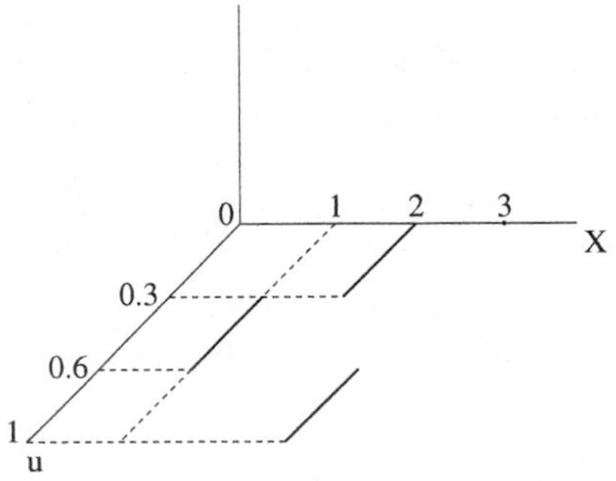

Fig. 2.23

Using previous notation we have $I_1 = [0.3, 0.6]$, $I_2 = [0, 0.3]$, and $I_3 = [0.6, 1]$. Standard operations may be defined using standard interval computations:

$$(\hat{A} \cup \hat{B})_x = [a_x^1 \vee b_x^1, a_x^2 \vee b_x^2],$$
$$(\hat{A} \cap \hat{B})_x = [a_x^1 \wedge b_x^1, a_x^2 \wedge b_x^2],$$
$$c(\hat{A})_x = [1 - a_x^2, 1 - a_x^1].$$

Here, $\hat{A}_x = [a_x^1, a_x^2]$, $\hat{B}_x = [b_x^1, b_x^2]$, and $c(\hat{A})$ denotes the complement of $\hat{A}$. Clearly, the family $\{A_x \,|\, x \in X\}$ determines the set $\hat{A}$.

Given a type 2 set $\hat{A}$, one can define a fuzzy set $\hat{A}_e$ by setting $\hat{A}_e(x) = u_x$, where u_x is a number selected from the interval I_x. The set $\hat{A}_e$ is said to be embedded in $\hat{A}$. If X and U are discretized and X has n elements, and I_{x_i} has M_i elements, the total number of embedded sets is $\prod_{i=1}^{n} M_i$.

We now consider a set of if-then rules where the antecedents and the consequents are fuzzy sets of type 2 and the input is a number denoted by x^*. First, we look at one rule:

If X is $\hat{A}$, then Y is $\hat{B}$,

where $\hat{A}$ and $\hat{B}$ are fuzzy sets of type 2. We assume, moreover, that all relevant spaces have been discretized. Let $\hat{A}_{e,h}$ and $\hat{B}_{e,j}$ be typical embedded sets in $\hat{A}$ and $\hat{B}$, as defined previously, $1 \le h \le n_A$, $1 \le j \le n_B$. We then consider the rule

If X is $\hat{A}_{e,h}$, then Y is $\hat{B}_{e,j}$.

There are $n_A n_B$ such rules. Given the numerical input x^*, each of these rules produces a fuzzy output, as explained in a previous subsection. Denote the output by $G_{h,j}$. In fact,

$$G_{h,j}(y) = \hat{A}_{e,h}(x^*) t \hat{B}_{e,j}(y),$$

where t denotes a t-norm. The output produced by the rule If X is $\hat{A}$, then Y is $\hat{B}$ is then $\{G_{h,j} \,|\, 1 \le h \le n_A, 1 \le j \le n_B\}$. Thus, if G denotes the output, we have

$$G(y) = \{G_{h,j}(y) \,|\, 1 \le h \le n_A, 1 \le j \le n_B\}.$$

Given any type 2 set $\hat{A}$, one can define $\underline{\hat{A}}$ and $\overline{\hat{A}}$ by $(\underline{\hat{A}})_x = a_x$ and $(\overline{\hat{A}})_x = b_x$, where $I_x = [a_x, b_x]$. $(\underline{\hat{A}})$ and $(\overline{\hat{A}})$ are the lower and the upper membership functions of $\hat{A}$, respectively. It can be shown that $\underline{G}$ and $\overline{G}$ are given by the outputs of the rules If X is $\underline{\hat{A}}$, then Y is $\underline{\hat{B}}$ and If X is $\overline{\hat{A}}$, then Y is $\overline{\hat{B}}$. Then, $G(y) = [\underline{G}(y), \overline{G}(y)]$. The formula can be naturally extended to multiple antecedents; no discretization required:

$$\underline{G}(y) = \left(T_{m=1}^{p} \underline{\hat{A}_m}(x_m^*)\right) t \underline{B}(y),$$

$$\overline{G}(y) = \left(T^{p}_{m=1}\overline{\hat{A}_m(x^*_m)}\right) t\overline{B}(y),$$
$$\hat{G}(y) = [\underline{G}(y), \overline{G}(y)].$$

Here, T and t denote t-norms that may or may not be the same. Finally, if there are N such rules, each rule produces an output G_l, as indicated earlier, and the set of N rules produces an output G that is an aggregation of G_l, $1 \leq l \leq N$. Typically,

$$\hat{G}(y) = \cap^{N}_{l=1}\hat{G}_l(y),$$

where the union is as defined earlier.

The Defuzzification Process

How is an interval valued function, such as G defuzzified? The process is accomplished in two steps: type reduction and defuzzification. Type reduction extracts a standard fuzzy set from G that is then defuzzified by any method indicated in a previous subsection. A number of methods leading to type reduction are available, just as a number of defuzzification methods that can be used. A very straightforward type reduction could be obtained by setting

$$T(y) = (\underline{G}(y) + \overline{G}(y))/2.$$

This is followed by one of the standard defuzzification methods applied to T.

We now briefly sketch one type of defuzzification: the height defuzzification. The k-th rule of type 2 produces an output as described earlier, $\hat{B}_k(y) = [\underline{B_k}(y), \overline{B_k}(y)]$, $1 \leq k \leq N$. Obtain the midpoint function $T_k(y) = (\underline{B_k}(y) + \overline{B_k}(y))/2$. Obtain y_k, a point at which T_k is maximum. If T_k is maximum on a set of points, take the average; see Figure 2.24.

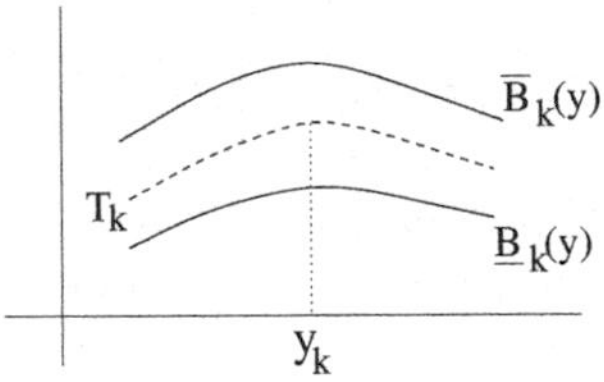

Fig. 2.24

Reorder the N rules if necessary so that we have $y_1 \leq y_2 \leq \cdots \leq y_N$. For each y_i, consider the interval I_{y_i} and partition that interval into M_i points. Consider the set

$$B_h(x) = \left\{ \sum_{i=1}^{N} y_i u_i / \sum_{i=1}^{N} u_i \mid u_i \in I_{y_i} \right\}.$$

A typical element in that set is the center of gravity of the function whose value at y_i is u_i. There are $\prod_{i=1}^{N} M_i$ such centers. It can be shown that $B_h(x)$ is actually an interval. Here, the notation $B_h(x)$ is used to stress that the formula that yields $B_h(x)$ depends on the input x. Thus, $B_h(x) = [c_l, c_r]$, where $c_l = \inf\{\sum y_i u_i / \sum u_i\}$ and $c_r = \sup\{\sum y_i u_i / \sum u_i\}$ over $u_i \in I_{y_i}$. The defuzzification is obtained by taking the midpoint of $[c_l, c_r]$. In the present context, type reduction refers to the defuzzification of fuzzy sets whose membership at y_i is u_i. That defuzzification produces the corresponding centers of gravity $\sum y_i u_i / \sum u_i$, where $u_i \in I_{y_i}$. The process in illustrated in Figure 2.25.

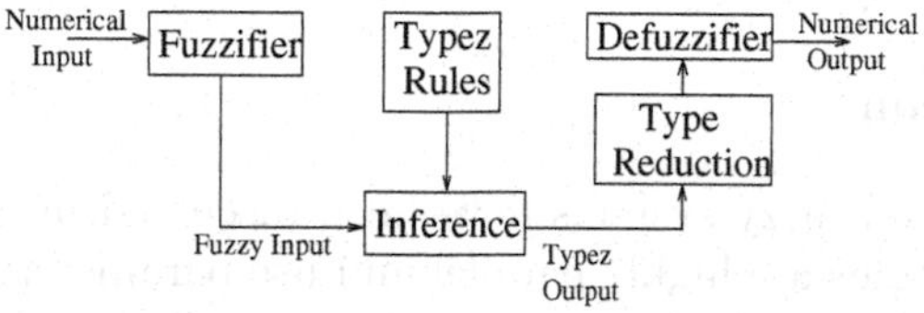

Fig. 2.25

Consider the rule If x is Low, then y is Average, where the membership functions of Low and Average are shown in Figure 2.26a. The idea is to factor in uncertainty due to the difference in expert opinions by replacing the above rule by If x is $\hat{\text{Low}}$, then y is $\hat{\text{Average}}$. Memberships of type 2 of $\hat{\text{Low}}$ and $\hat{\text{Average}}$ are shown in Figure 2.26b.

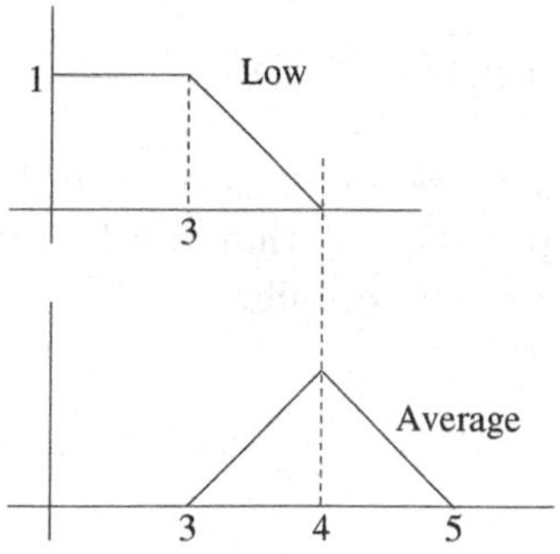

Fig. 2.26a

For more information on fuzzy sets of type 2 see [28], [29], and [30].

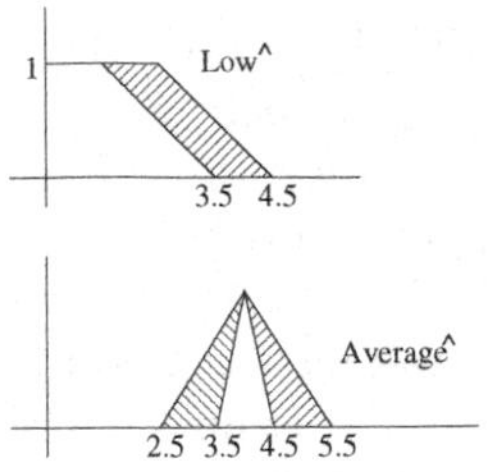

Fig. 2.26b

2.3 Neuro-Fuzzy Systems

2.3.1 Introduction

Neural networks and fuzzy systems have been discussed in the two previous sections. Both systems are highly parallel and use parametric representations for weights and for membership functions. In neural nets, knowledge representation and extraction is a difficult process. In a fuzzy system, knowledge representation takes the natural form of if-then statements. On the other hand, the success of fuzzy systems depends on how accurate the membership functions are. In addition, it is not always clear how to translate knowledge into if-then rules. It is therefore natural to seek a hybrid technology that would combine the advantage of neural nets (i.e., their capability to adapt to a given environment with the advantage of fuzzy systems i.e., their natural knowledge representation). The TSK model has been defined in a previous subsection. Recall that a typical rule has the form

If X is A and Y is B, then $Z = f(x, y)$.

Here, A and B denote fuzzy sets and f is often either a constant or a first-order polynomial. It was pointed out that no defuzzification was necessary. The output is given by the single formula

$$C(x_1, \ldots, c_n) = \sum_{i=1}^{N} \bar{w}_i f_i(x_1, \ldots, x_n)$$

if there are N rules and if f_i is the consequent of the i-th rule. Here, $\bar{w}_i$ denote the normalized weights

$$\bar{w}_i = w_i / \sum_{j=1}^{N} w_j,$$

where w_i is the strength of the i-th rule when the input is $(x_1, \ldots, x_n)^T$.

2.3.2 Fuzzy Neurons

We now extend the idea of a neuron to define "fuzzy neuron." Although a general definition could be given, we mainly will use three types of neuron. The first type has a membership function stored that it uses to weight a fuzzy input, whereas the other two types use boolean operations. The three types are shown in Figures 2.27a – 2.27c respectively.

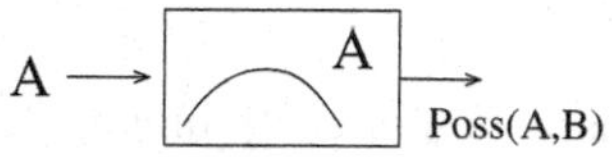

Fig. 2.27a

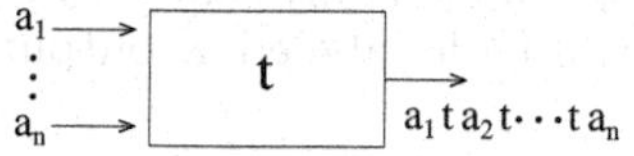

Fig. 2.27b

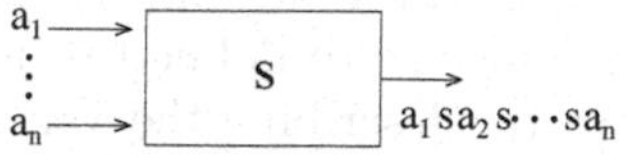

Fig. 2.27c

The input to the first type is a fuzzy set B and the output is Poss(A, B). In particular, if the input is a number x, then the output is $A(x)$. The other two types perform a t-norm and an s-norm operation, respectively. Inputs here are numbers in $[0, 1]$. Often t is the minimum or the product operation and s is often the maximum operation.

2.3.3 The Adaptive Neuro-Fuzzy Inference System (ANFIS)

It is straightforward to convert TSK-type rules into an ANFIS architecture. For example, suppose that we have the following two rules

R_1: If x is A_1 and y is B_1, then z is $f_1(x, y)$.

R_1: If x is A_2 and y is B_2, then z is $f_2(x, y)$.

Here, $f_1(x, y) = p_1x + q_1y + r_1$ and $f_2(x, y) = p_2x + q_2y + r_2$. This system of rules translates into a five-layer ANFIS shown in Figure 2.28.

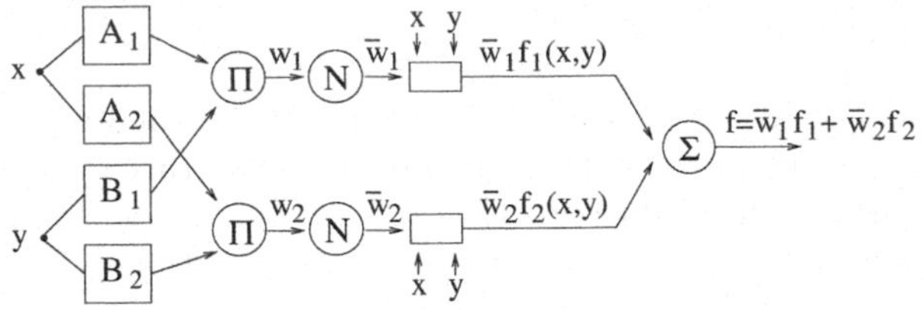

Fig. 2.28

In this case, the input is a numerical pair (x, y). The first layer consists of fuzzy neurons that output $A_1(x)$, $A_2(x)$, $B_1(y)$, and $B_2(y)$. The second layer, whose nodes are labeled Π, outputs the t products. For example, it could output $A_1(x)B_1(y)$ and $A_2(x)B_2(y)$. The third layer, whose nodes are labeled N, simply normalizes the previous strengths of the two rules. The next layer simply outputs $f_1(x, y)$ and $f_2(x, y)$ scaled by the normalized strengths. Finally, the last layer's single node labeled Σ outputs what the TSK system would output.

2.3.4 A Comparison Among Three Approaches

In Figure 2.21 we have shown a simple vehicle whose goal is to avoid obstacles. We have stated the type of fuzzy rules that may control this vehicle. A purely neural net approach to this could be to define two neural nets for which the input is the 4-dimensional vector describing the four distances to the obstacle $(\text{Left}, \text{Front}, \text{Right}, \text{Rear})^T$ and the output would be the corresponding forces applied to the left motor and to the right motor. Thus, the two neural nets would connect the four sensors input to the two motors. The weight could initially be defined by, for example, the matrix W.

	Left	Front	Right	Rear	Bias
A	0.2	−0.5	−0.3	0.2	0.3
B	−0.4	0.1	0.2	0.3	0.3

At every collision, the matrix W should be updated, using, for example, backpropagation. The target would be provided by the desired trajectory before collision has occurred. It is certainly not clear, by looking at the weights, how to characterize the behavior of the vehicle. That characterization is very apparent using the fuzzy approach.

Consider now the neuro-fuzzy approach. Rules could take the following forms:

R_1: If Left Distance is Large and Front Distance is Large, then Force applied to Left Motor is $f_1(x, y)$.

R_2: If Front Distance is Short, then Force applied to Left Motor is $f_2(y)$.

Here, x is the Left distance and y is the Front distance: $f_1(x, y) = p_1x+q_1y+r_1$ and $f_2(y) = q_2y+r_2$. What is needed is a learning algorithm to estimate p_1, q_1, r_1, q_2, and r_2 as well as the parameters involved in the membership functions of Large and Short.

Learning Algorithm for ANFIS

We need a learning algorithm for systems such as the one sketched in Figure 2.28. This system could be viewed as an example of an adaptive net and its parameters estimated as described in section 2.1. Perhaps a better way to proceed is to partition the parameter set into the consequent parameters – those coming from

$$f_i(x, y, z, \ldots) = p_{i,1}x + p_{i,2}y + p_{i,3}z + \cdots + r_i$$

– and the antecedent parameters, i.e. those coming from the defining membership functions of $A_{i,1}, \ldots, A_{i,p}$, assuming that there are p variables in the antecedent. We note that the output $f = \sum \bar{w}_i f_i$ is a linear function in $p_{i,1}$, $p_{i,2}$, $p_{i,3}$,$\ldots$,r_i, $1 \leq i \leq N$, where N is the number of rules. These parameters can therefore be estimated by a least squares fit. The antecedent parameters can be estimated by using the backpropagation algorithm for adaptive nodes.

2.3.5 Organizing Rules and Architectures

Consider the four rules:

If x is A_1 and y is B_1, then $z_1 = c_{1,1}$.

If x is A_2 and y is B_2, then $z_1 = c_{2,1}$.

If x is D_1 and y is E_1 then $z_2 = c_{1,2}$.

If x is D_2 and y is E_2, then $z_2 = c_{2,2}$.

Here, $c_{i,k}$ refers to the formula giving z_k in the i-th rule and

$$c_{i,k} = p_{i,k}x + q_{i,k}y + r_{i,k}.$$

The extension of this notation to the case of more than two variables is obvious. The diagram in Figure 2.29 shows a possible structure for these four rules. The idea is to put together in parallel ANFIS structures for each output. Here, O_1 and O_2 obviously refer to the outputs z_1 and z_2 generated by these four rules when the input is the numerical vector $(x, y)^T$. Such a structure is called "Coactive Adaptive Neuro-Fuzzy Inference System" or CANFIS in the

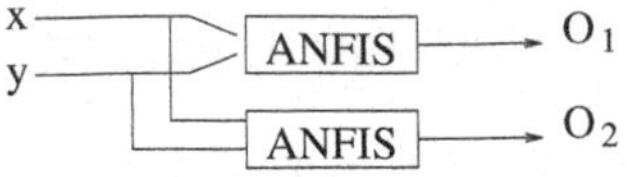

Fig. 2.29

literature. Thus, one way to construct a CANFIS is to put several ANFIS in parallel (one ANFIS for each output). This way of proceeding is often inefficient since no parameter sharing takes place and one is typically faced with a large number of parameters to estimate. A better way to proceed is to use multiple consequents. For example, consider the following two rules:

If x is A_1 and y is B_1, then $z_1 = c_{1,1}$ and $z_2 = c_{1,2}$,

If x is A_2 and y is B_2, then $z_1 = c_{2,1}$ and $z_2 = c_{2,2}$.

The diagram in Figure 2.30 shows a possible structure for these two rules.

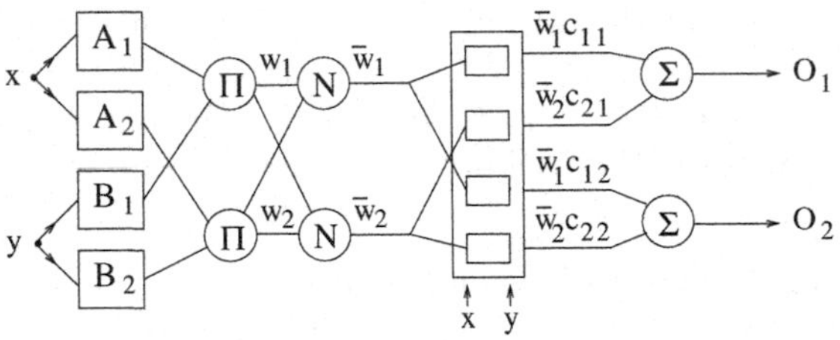

Fig. 2.30

Clearly, in this approach, parameter sharing takes place and fewer parameters are to be estimated for this CANFIS. On can draw an interesting comparison between a CANFIS and a standard neural net. The part following the N layer in Figure 2.30 could be replaced by the system shown in Figure 2.31.

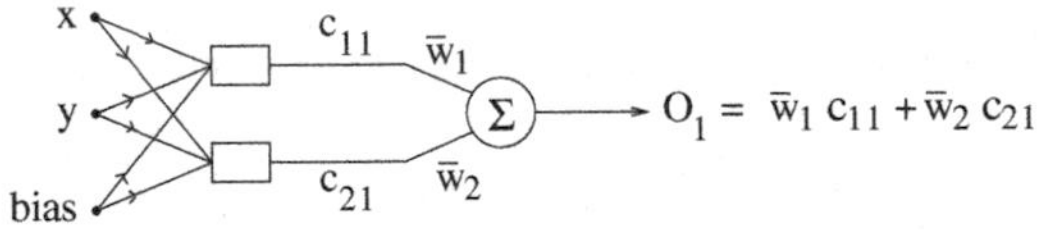

Fig. 2.31

Here, the weights connecting x, y, and the bias to the first neuron of the first layer are $p_{1,1}$, $q_{1,1}$, and $r_{1,1}$, whereas the weights connecting x, y, and the bias to the second neuron are $p_{2,1}$, $q_{2,1}$, and $r_{2,1}$. Thus, the two neurons will produce $c_{1,1}$ and $c_{2,1}$ as outputs. The weights connecting these to the Σ node are $\bar{w}_1$ and $\bar{w}_2$, obtained as outputs from the first part of the system shown in Figure 2.30. Here, the system produces the output O_1. A similar net, not shown in the diagram, is needed to produce the second output O_2. Training of this set will determine the weights $p_{i,j}$, $q_{i,j}$, and $r_{i,j}$. From this it is therefore clear that CANFIS can be viewed as appending the linguistic component in front of a standard net. In fact, the neural component may be modified to handle nonlinear functions of x and y. Consider the general m-th rule:

R_m: If x_1 is A_1 and x_2 is A_2 and ..., then $z_1 = c_{m,1}$, $z_2 = c_{m,2}, \cdots, z_p = c_{m,p}$.

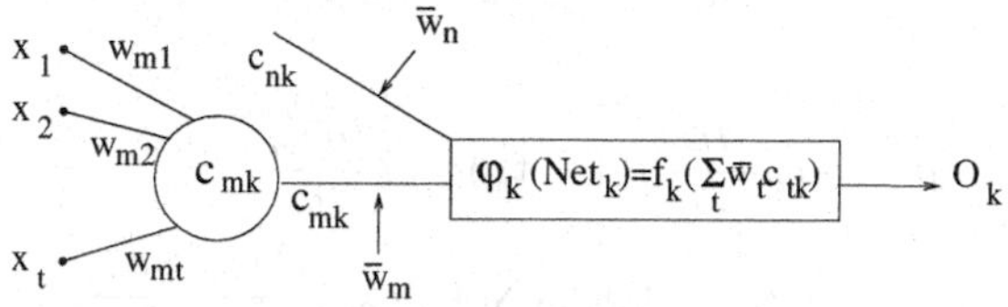

Fig. 2.32

Figure 2.32 shows how a set of rules of type R_m generate the outputs O_k.

We have a training set and the weights $w_{m,k}$, $1 \leq k \leq t$ are adjusted for $1 \leq m \leq N$ if we have N rules of type R_m. In the case discussed previously, f was the identity function. Here, f_k could, for example, be $f_k(u) = 1/(1+e^{-u})$. The net prior to the $\varphi_k(\text{Net}_k)$ box corresponds to the action of one rule (R_m in this case) and could, in principle, be trained by itself. The $\varphi_k(\text{Net}_k)$ box plays the role of consensus builder by weighting every rule according to its normalized strength. One could still generalize this by allowing the transfer function at $c_{m,k}$ to be other than the identity function.

2.3.6 Updating the Consequents and the Antecedents

In order to update the consequents, one needs to update the weights $w_{m,k}$, $1 \leq k \leq t$, $1 \leq m \leq N$. In order to do this, the function

$$E = (t_k - O_k)^2/2$$

needs to be minimized, where t_k is the corresponding target and $O_k = \varphi_k(\text{Net}_k)$. We apply the steepest descend method: $\Delta w_{m,i} = -\eta_m \partial E / \partial w_{m,i}$, where η_m is a small positive number. So,

$$\Delta w_{m,i} = -\eta_m \frac{\partial E}{\partial \text{Net}_k} \frac{\partial \text{Net}_k}{\partial w_{m,i}}$$

$$
\begin{aligned}
&= -\eta_m \frac{\partial E}{\partial O_k} \frac{\partial O_k}{\partial \mathrm{Net}_k} \frac{\partial \mathrm{Net}_k}{\partial c_{m,k}} \frac{\partial c_{m,k}}{\partial w_{m,i}} \\
&= -\eta_m \frac{\partial E}{\partial O_k} \varphi_k'(\mathrm{Net}_k) \bar{w}_m x_i \\
&= \eta_m (t_k - O_k) f_k'(\mathrm{Net}_k) \bar{w}_m x_i .
\end{aligned}
$$

Let a be a parameter involved in the antecedent. We have $\triangle a = -\eta_a \partial E / \partial a$, where η_a is a small positive number. Then

$$
\triangle a = -\eta_a \frac{\partial E}{\partial O_k} \frac{\partial O_k}{\partial a} = -\eta_a \frac{\partial E}{\partial O_k} \frac{\partial O_k}{\partial \mathrm{Net}_k} \frac{\partial \mathrm{Net}_k}{\partial w_m} \frac{\partial w_m}{\partial a} .
$$

Since

$$
\frac{\partial \mathrm{Net}_k}{\partial w_m} = \frac{\partial \mathrm{Net}_k}{\partial \bar{w}_m} \frac{\partial \bar{w}_m}{\partial w_m},
$$

it follows that

$$
\begin{aligned}
\triangle a &= -\eta_a \frac{\partial E}{\partial O_k} \varphi_k'(\mathrm{Net}_k) c_{m,k} \frac{\partial \bar{w}_m}{\partial a} \\
&= \eta_a (t_k - O_k) \varphi_k'(\mathrm{Net}_k) c_{m,k} \frac{\partial \bar{w}_m}{\partial w_m} \frac{\partial w_m}{\partial a} .
\end{aligned}
$$

Recall that w_m denotes the strength of the m-th rule that was defined by

$$
w_m = A_1(x_1) t A_2(x_2) t \cdots t A_l(x_l),
$$

where t denotes a t-norm. Since a is a parameter involved in the membership functions, w_m is a function of a. Also, $\bar{w}_m = w_m / \sum w_i$.

For further readings on neuro-fuzzy systems, the reader is referred to the works in [13], [14], [15], [31], [32], [40], and [41].

2.4 The Theory of Evidence

2.4.1 Introduction

There are three types of uncertainty. The fuzziness comes about from the absence of sharp boundaries separating classes of objects. The nonspecificity comes about from the size of the set of alternatives and the strife comes about from the conflicts among the available alternatives. Entropy measures fuzziness and is more or less a direct extension of entropy as known in probability theory. If p_i, $1 \le i \le n$, is a discrete probability distribution, the entropy function is defines as

$$
H(p_1, \ldots, p_n) = -\sum p_i \log_2 p_i .
$$

This function is zero when $p_{i^*} = 1$ and $p_j = 0$, $j \neq i^*$ (i.e., when we are certain that alternative i^* will occur, and can be shown to reach a maximum when $p_i = 1/n$ for all i (i.e., when all outcomes have equal probabilities).

If A is a fuzzy set defined on a finite universe of discourse and $A = \sum \alpha_i / x_i$ with all the x_i distinct, then the entropy is defined by

$$H(A) = \sum h(A(x_i)),$$

where h is a function from $[0, 1]$ to $[0, 1]$ satisfying the following properties:

$h(x) = 0$ if and only if $x = 0$ or $x = 1$.

$h(x)$ is maximum at $x = 1/2$.

$h(x) = h(1 - x)$.

h is monotone increasing on $[0, 1/2]$ and monotone decreasing on $[1/2, 1]$.

Examples of such functions are $h(x) = 1 - |2x - 1|$, $h(x) = 4x(1 - x)$, and $h(x) = -x \log_2 x - (1 - x) \log_2(1 - x)$.

It follows that $H(A)$ is maximum at $A(x) = 1/2$ for all x; that is H is maximum at the "maximal fuzzy set." Also, $H(A) = 0$ if $A(x) \in \{0, 1\}$ for all x; that is H is 0 on standard nonfuzzy sets. Let $A_{1/2}$ denote the standard non-fuzzy set $\{x \mid A(x) \geq 1/2\}$. If $A_{1/2}$ denotes the membership in that set and if $h(x) = 1 - |2x - 1|$, it can be shown that

$$H(A) = 2 \sum_i |A(x_i) - A_{1/2}(x_i)|.$$

So, $H(A)$ is twice the Hamming distance of A to $A_{1/2}$. Intuitively, the more A is removed from the standard set $A_{1/2}$, the higher the entropy of A is and the more fuzzy A is.

In the continuous case, the entropy H can be defined as

$$H(A) = \int h(A(x))\, dx.$$

2.4.2 Evidence Theory

The concepts of non specificity and strife are best defined in the general context of evidence theory.

Definition 1. *Let $\mathcal{A}$ be the nonempty family of subsets of the set X. A "fuzzy measure" on $< X, \mathcal{A} >$ is a function g from $\mathcal{A}$ into $[0, 1]$ such that the following hold:*

(1) $g(\emptyset) = 0$ *and* $g(X) = 1$.

(2) If $A \subset B \subset X$ *then* $g(A) \leq g(B)$.

(3) If A_n is an increasing sequence of subsets in $\mathcal{A}$ (i.e., $A_n \subset A_{n+1}$ for every n), and if $\cup A_n \in \mathcal{A}$, then $g(\cup A_n) = \lim g(A_n)$.

(4) If A_n is a decreasing sequence of subsets in $\mathcal{A}$ and $\cap A_n \in$ A, then $g(\cap A_n) = \lim g(A_n)$.

Of course, if X is a finite set, then properties (3) and (4) are vacuous. The interpretation of $g(A)$ is "the total evidence" that some unknown element belongs to the set A. From the definition it follows that

$$g(A \cap B) \leq g(A) \wedge g(B) \quad \text{and} \quad g(A \cup B) \geq g(A) \vee g(B).$$

Note that a probability function is a special case of a fuzzy measure. At times, properties (1), (2), and (3) will be the only properties assumed, and sometimes properties (1), (2), and (4) will be the only properties assumed. Two fuzzy measures are of particular importance: the belief and the plausibility. These measures will be denoted by Bel and Pls.

Definition 2. *Bel is defined as a function from the subsets of X to $[0,1]$ with the following properties:*

(1) $\mathrm{Bel}(\emptyset) = 0$ *and* $\mathrm{Bel}(X) = 1$.

(2) $\mathrm{Bel}(A_1 \cup \cdots \cup A_n) \geq \sum_j \mathrm{Bel}(A_j) - \sum_{j<k} \mathrm{Bel}(A_j \cap A_k) + \cdots + (-1)^{n+1}\mathrm{Bel}(A_1 \cap \cdots \cap A_n)$.

If X is infinite we also require property (4) of fuzzy measures. Note that $\mathrm{Bel}(A) + \mathrm{Bel}(\bar{A}) \leq 1$.

Pls is also defined as a function from subsets of X to $[0,1]$ satisfying the following properties:

(1) $\mathrm{Pls}(\emptyset) = 0$ *and* $\mathrm{Pls}(X) = 1$.

(2) $\mathrm{Pls}(A_1 \cap \cdots \cap A_n) \leq \sum_j \mathrm{Pls}(A_j) - \sum_{j<k} \mathrm{Pls}(A_j \cup A_k) + \cdots + (-1)^{n+1}\mathrm{Pls}(A_1 \cup \cdots \cup A_n)$.

If X is infinite, we also require property (3) of fuzzy measures. Note that $\mathrm{Pls}(A) + \mathrm{Pls}(\bar{A}) \geq 1$.

Both Bel and Pls can be characterized by a third function called mass.

Definition 3. *A mass m is a function from subsets of X to $[0,1]$ satisfying*

$$m(\emptyset) = 0 \quad \textit{and} \quad \sum_{A \subset X} m(A) = 1.$$

It can be shown that if

$$h_1(A) = \sum_{B \subset A} m(B) \quad \text{and} \quad h_2(A) = \sum_{B \cap A \neq \emptyset} m(B),$$

then h_1 is a belief function and h_2 is a plausibility function. The mass m is said to generate belief h_1 and plausibility h_2. Conversely, we have the following:

If Bel is any belief function, then

$$m(A) = \sum_{B \subset A} (-1)^{|A-B|} \text{Bel}(B)$$

is a mass function that generates belief Bel.

If Pls is any plausibility function, then

$$m(A) = \sum_{B \subset A} (-1)^{|A-B|} (1 - Pls(\bar{B}))$$

is a mass function that generates plausibility Pls.

Whereas, as pointed out earlier, a fuzzy measure applied to a set A measures the total evidence that a partially known element belongs to A, $m(A)$ measures the evidence (not total) that a partially known object is precisely in A ($m(A)$ does not include, for example, any evidence that the object might be in some subset of A). In this connection, it should be stressed that $A \subset B$ does not necessarily imply $m(A) \leq m(B)$. For example, if we have three types of aircraft: bomber (B), fighter (F), and passenger plane (PP) – and we are almost sure that the aircraft we are focusing on is a military plane, the corresponding mass defined on $X = \{\text{B}, \text{F}, \text{PP}\}$ could be

$$m(\{\text{B}, \text{F}\}) = 0.9, \qquad m(\{\text{B}, \text{F}, \text{PP}\}) = 0.1.$$

The reason m is small on $\{\text{B}, \text{F}, \text{PP}\}$ is that evidence is strong the aircraft is not any one of the three aircraft since PP is practically ruled out. On the other hand, the total evidence for X is

$$\text{Bel}(\{\text{B}, \text{F}, \text{PP}\}) = m(\{\text{B}, \text{F}\}) + m(\{\text{B}, \text{F}, \text{PP}\}) = 1.$$

2.4.3 Composition of Masses

Given a mass m, those subsets of X on which m is not zero are called the focal elements of m. If m_1 and m_2 denote masses generated by independent sources, then the composition of m_1 and m_2 is defined to be a mass whose focal elements are the intersections of the focal elements of m_1 and m_2. The composition mass is defined by

$$(m_1 \oplus m_2)(A) = \sum_{B \cup C = A} m_1(B)m_2(C) / \sum_{B \cap C \neq \emptyset} m_1(B)m_2(C).$$

In other words, we discount conflicting focal elements. We can rewrite the above as follows:

$$(m_1 \oplus m_2)(A) = \sum_{B \cup C = A} m_1(B)m_2(C)/(1 - K),$$

where

$$K = \sum_{B \cap C = \emptyset} m_1(B)m_2(C).$$

One can define $m_1 \oplus \cdots \oplus m_n$ inductively by

$$m_1 \oplus \cdots \oplus m_n = (m_1 \oplus \cdots \oplus m_{n-1}) \oplus m_n.$$

Example 1. Assume we have three persons who are suspected of some crime. The three persons are John, James, and Jane. James and Jane are siblings. There is a piece of evidence that indicates a male was involved in the crime and the strength of the evidence (on a scale of 0 to 1) is 0.7. There is a slight amount of evidence that Jane is actually the guilty one and the strength of that evidence is 0.2. A totally different piece of evidence seems to indicate that siblings were involved and the strength of that is 0.6. That second piece of evidence seems to confirm what the previous evidence had shown: a slight possibility that Jane was involved. The strength to support this assumption is still 0.2. Putting these two pieces together, how strong is the evidence against any single individual? Let m_1 and m_2 be the masses corresponding to the two pieces of evidence. We have

$$m_1\{\text{John, James}\} = 0.7, \qquad m_1\{\text{Jane}\} = 0.2.$$

There is a floating .1 that we assign to the whole set as that .1 could be of concern to any one of the individuals, so

$$m_1\{\text{John, James, Jane}\} = 0.1.$$

Similarly,

$$m_2\{\text{James, Jane}\} = 0.6, \quad m_2\{\text{Jane}\} = 0.2, \quad m_2\{\text{John, James, Jane}\} = 0.2.$$

We put the two pieces together by combining m_1 and m_2. Since the focal elements of $m_1 \oplus m_2$ are the intersections of the focal elements of m_1 and m_2 and since {John} is not an intersection of two such focal elements, $(m_1 \oplus m_2)\{\text{John}\} = 0$. So, there is no evidence pointing to John being the sole guilty person. Let

$$A = \{\text{John, James}\}, \quad B = \{\text{Jane}\}, \quad C = \{\text{James, Jane}\},$$

and $U = \{\text{John, James, Jane}\}$. Then, $K = m_1(A)m_2(B) = 0.14, 1-K = 0.86$ and

$$(m_1 \oplus m_2)(A) = m_1(A)m_2(U)/(1-K) = 0.14/0.86 \approx 0.16.$$

Thus, although there is no evidence against John alone, there is approximately 0.16 evidence that John or James did commit the crime, but we cannot determine which of those two. Next, we have

$$(m_1 \oplus m_2)(\{\text{James}\}) = m_1(A)m_2(C)/(1-K) = 0.42/0.86 \approx 0.49.$$

Although none of the two pieces of evidence pointed to James acting alone, pulling all the information together yields an evidence of approximately 0.49 that James was the sole perpetrator. Finally,

$$\begin{aligned}
&(m_1 \oplus m_2)\{\text{Jane}\} = \\
&(m_1(B)m_2(U) + m_1(B)m_2(b) + m_1(B)m_2(U)) + m_1(U)m_2(B))/(1-K) \\
&= 0.22/0.86.
\end{aligned}$$

Whereas the separate pieces of evidence gave 0.2 evidence against Jane, the combined evidence against Jane is slightly higher, approximately 0.26. Obviously, this type of reasoning can be used in object recognition problems where features are known with specified probabilities.

2.4.4 Possibility Theory

A special case of evidence theory is provided by possibility theory. Here, the mass has nested focal elements. Thus, if a mass has focal elements $A_1, \cdots, A_n$, we have $A_1 \subset \cdots \subset A_n$. If this is so, it is straightforward to show that

$$\text{Bel}(A \cap B) = \min\{\text{Bel}(A), \text{Bel}(B)\}$$

and

$$\text{Pls}(A \cup B) = \max\{\text{Pls}(A), \text{Pls}(B)\}.$$

In this case, one often uses the terms necessity (Nec) and possibility (Pos) instead of belief and plausibility. Again, it is straightforward to show that

$$\begin{aligned}
&\text{Nec}(A) = 1 - \text{Pos}(\bar{A}), \\
&\text{Nec}(A) > 0 \quad \text{implies} \quad \text{Pos}(A) = 1, \\
&\text{Pos}(A) < 1 \quad \text{implies} \quad \text{Nec}(A) = 0.
\end{aligned}$$

The following is an important theorem that we state without a proof.

Theorem 1. *Every possibility function is uniquely determined by a possibility distribution function* $r : X \to [0,1]$ *so that* $\text{Pos}(A) = \sup\{r(x) \mid x \in A\}$. *The function* r *is defined by* $r(x) = \text{Pos}\{x\}$. *If, in addition,* X *is discrete (i.e.,* $X = \{x_1, \ldots, x_n\}$*), then* $1 = r_1 \geq r_2 \geq \cdots \geq r_{n+1} = 0$ *and* $m(A_i) = r_i - r_{i+1}$ *where* $A_i = \{x_1, \ldots, x_i\}$ *and* $r_i = r(x_i) = \text{Pos}\{x_i\}$.

Example 2. Let $X = \{x_1, \ldots, x_7\}$. Let m be a mass on the subsets of X whose focal elements are $A_2 = \{x_1, x_2\}$, $A_3 = \{x_1, x_2, x_3\}$, $A_6 = \{x_1, \ldots, x_6\}$, and $A_7 = X$. Note that the focal elements of m are nested: $A_2 \subset A_3 \subset A_6 \subset A_7$. Assume that

$$m(A_2) = 0.3, \quad m(A_3) = 0.4, \quad m(A_6) = 0.1, \quad m(A_7) = 0.2.$$

Then since $r_i = \text{Pos}\{x_i\} = \sum_{j=i}^{7} m(A_j)$, we obtain $r_1 = 1$, $r_2 = 1$, $r_3 = 0.7$, $r_4 = 0.3$, $r_5 = 0.3$, $r_6 = 0.3$, and $r_7 = 0.2$. Now, for example,

$$\text{Poss}\{x_3, x_4, x_5\} = \sup\{r_3, r_4, r_5\} = 0.7.$$

2.4.5 Relation of Possibility to Fuzzy Sets

Let F be a fuzzy subset of X. If F is used as a possibility distribution function, taking into account the discussion in the previous section, it is natural to define a possibility associated with F by

$$\text{Pos}_F(A) = \sup_{x \in A} F(x).$$

It is also natural to define

$$\text{Nec}_F(A) = 1 - \text{Pos}_F(\bar{A}).$$

Note that when A is a standard crisp set, then (see Section 2.4)

$$\text{Pos}_F(A) = \sup_x (A(x) \wedge F(x)) = \text{Poss}(A, F).$$

Example 3. Let F be a fuzzy set defined by $F = 0.2/1 + 0.4/2 + 0.5/3 + 1/4 + 0.5/5 + 0.4/6 + 0.2/7$. We now order the elements of X in decreasing order of their memberships in F. We then have the list 4, 3, 5, 2, 6, 1, 7. The sets A_i are: $A_1 = \{4\}$, $A_2 = \{4, 3\}$, $A_3 = \{4, 3, 5\}$, $A_4 = \{4, 3, 5, 2\}$, $A_5 = \{4, 3, 5, 2, 6\}$, $A_6 = \{4, 3, 5, 2, 6, 1\}$, and $A_7 = \{4, 3, 5, 2, 6, 1, 7\}$. Using the formula $m(A_i) = r_i - r_{i+1}$, where $r_i = F(x_i)$ (with $r_1 = 1$ and $r_8 = 0$), we see that the focal elements of m are A_1, A_3, A_5, and A_7. Then, for example,

$$\text{Poss}\{x_2, x_5, x_7\} = \max\{r_2, r_5, r_7\} = \max\{F(2), F(5), F(7)\} = 0.5.$$

On the other hand, $m(A_1) = 0.5$, $m(A_3) = 0.1$, $m(A_5) = 0.2$, and $m(A_7) = 0.2$, so

$$\text{Pls}\{x_2, x_5, x_7\} = \sum_{j=2}^{7} m(A_j) = .5.$$

The two computations yield the same value.

It is clear that if focal elements are single points, the mass is then a probability distribution and Bel = Pls and coincide with probability measures. Thus, in general, Bel < Pls and can be viewed as the lower and the upper bound of some unknown probability measure. In some sense, Pls − Bel is a measure of the uncertainty regarding the determination of some appropriate probability measure.

2.4.6 Nonspecificity

Let A be a standard crisp set. The nonspecificity of A is defined as

$$N(A) = \log_2 |A|,$$

where, of course, $|A|$ denotes the cardinality of A. Intuitively, this is how many splittings of A into equal halves is required to arrive at a single answer. Clearly, the larger the cardinality of A is, the higher the nonspecificity is. Another interpretation is the number of bits required to represent all of the elements of A. If A represents the set of currently viable alternatives and if new information comes in that reduces the number of viable alternatives to a set B where $B \subset A$, then the reduction of uncertainty is given by $N(A) - N(B) = \log_2(A/B)$. This could be used as a measure of the amount of information received, since the information received could reasonably be identified with the reduction in uncertainty.

How can this be generalized to the case where A is a fuzzy set? We define

$$N(A) = \frac{1}{h(A)} \int_0^{h(A)} \log_2 |A_\alpha|\, d\alpha.$$

Here, $h(A)$ denotes the height of A (i.e., $h(A) = \sup_x A(x)$). Furthermore, $|A_\alpha|$ denotes the α-cut of A; that is A_α is a crisp set defined by $A_\alpha = \{x \mid A(x) \geq \alpha\}$ and $|A_\alpha|$ denotes the length of A_α (assuming A is such that A_α is an interval for every $\alpha \geq 0$).

The obvious analog for the discrete case is

$$N(A) = \frac{1}{h(A)} \sum_{i=0}^{n-1} (\log_2 |A_{\alpha_i}|)(\alpha_{i+1} - \alpha_i),$$

where $A = \sum_{i=1}^{n} \alpha_i / x_i$, with $\alpha_n \geq \cdots \geq \alpha_1 \geq \alpha_0 = 0$. Clearly, if A is a crisp set, then $h(A) = 1$, $1 = \alpha_1 \geq \alpha_0 = 0$, $|A_\alpha| = |A|$, and we get back the nonspecificity formula for crisp sets.

If we have a possibility distribution $r =< r_1, \ldots, r_n >$ with $1 = r_1 \geq \cdots \geq r_n \geq r_{n+1} = 0$, the nonspecificity of r is defined by

$$N(r) = \sum_{i=2}^{n} (r_i - r_{i+1}) \log_2 i.$$

The reason $N(r)$ is defined as above is the following: Let F be a fuzzy set. Denote by r_F a possibility distribution associated with F to be defined later. Then an interesting result is

$$N(r_F) = N(F),$$

provided these quantities are redefined in the context of the theory of evidence. Given a mass m, the nonspecificity of m is defined as

$$N(m) = \sum_{A \mid m(A) \neq 0} m(A) \log_2 |A|,$$

that is, the sum is over all focal elements of m. Thus, $N(m)$ is the averaged nonspecificity over all focal elements of m. Now, given a fuzzy set F over a finite universe of discourse X, a mass m_F is generated that has focal elements $A_i = \{x_1, \ldots, x_i\}$, $i = 1, \ldots, n$, and a possibility distribution r_F is generated with $r_F(x_i) = \text{Pos}_F(x_i)$ and $m_F(A_i) = r_i - r_{i+1}$ (see Section 4.4). Let $r_{F,i} = r_F(x_i)$. Clearly, $|A_i| = i$. Then

$$N(m_F) = \sum_{i=1}^{n} m(A_i) \log_2 |A_i| = \sum_{i=2}^{n} (r_{F,i} - r_{F,i+1}) \log_2 i = N(r_F).$$

We thus define the nonspecificity of a fuzzy set F by

$$N(F) = N(m_F) = N(r_F).$$

Note that if m is a probability distribution, its focal elements are singletons and m, and in that case, since $\log_2 1 = 0$, we have $N(m) = 0$. Thus, a probability distribution has zero nonspecificity .

2.4.7 Strife

The strife function will generalize the Shanon entropy function and is again best defined in the context of the theory of evidence. We start by defining conflict over a subset A of X as

$$\text{Con}(A) = \sum_{B} m(B) \frac{|A - B|}{|A|}.$$

Note that if $A \subset B$, then $|A - B| = 0$; that is sets that contain A are not counted in the conflict because $A \subset B$ implies $A \to B$ (A implies B). So if B follows from A, it should not contribute to the conflict generated by A. We would like the strife function to generalize the concept of entropy function so we start by looking at the special case where m is a probability distribution p. In this case,

$$\text{Con}(\{x\}) = \sum_{y \neq x} p(y) = 1 - p(x).$$

So for the probability distribution the entropy is

$$-\sum_{x} p(x) \log_2 p(x) = -\sum_{x} p(x) \log_2 (1 - \text{Con}(\{x\})).$$

Then it is natural to define the strife function, relative to a mass m by

$$S(m) = -\sum_{A} m(A) \log_2 (1 - \text{Con}(A)),$$

where the sum is, of course, over the focal sets of m. Note that as $\mathrm{Con}(A)$ increases from 0 to 1, $-\log_2(1-\mathrm{Con}(A))$ increases from 0 to ∞. Also, note that $\mathrm{Con}(A)$ is the averaged proportion of elements of A that fail to be in B where B is a focal subset of A. The strife $S(m)$ can be rewritten as

$$\begin{aligned} S(m) &= -\sum_A m(A)\log_2[\sum_B m(B)(1-|A-B|/|A|)] \\ &= -\sum_A m(A)\log_2[\sum_B m(B)|A\cap B|/|A|] \\ &= \sum_A m(A)\log_2|A| - \sum_A m(A)\log_2[\sum_B m(B)|A\cap B|]. \end{aligned}$$

If we denote by $R(m)$ the second sum on the right-hand side, we have

$$S(m) = N(m) - R(m).$$

In particular, we can now define strife in the context of possibility theory, and therefore in the context of fuzzy sets. Recall that in the context of possibility theory we have nested focal elements $A_1 \subset \cdots \subset A_n$, where $A_i = \{x_1, \cdots, x_i\}$, $r_i = r(x_i)$, $1 = r_1 \geq \cdots \geq r_n \geq r_{n+1} = 0$ with $m(A_i) = r_i - r_{i+1}$. So in this case,

$$S(m) = N(m) - R(m) = \sum_{i=2}^{n}(r_i - r_{i+1})\log_2 i - R(m).$$

The quantity $R(m)$ can be written as

$$R(m) = \sum_{i+1}^{n}(r_i - r_{i+1})\log_2\left(\sum_{j=1}^{n}(r_j - r_{j+1})\min\{j,i\}\right).$$

So the final form of $S(m)$ is

$$S(m) = N(m) - \sum_{i=1}^{n}(r_i - r_{i+1})\log_2\left(i/\sum_{j=1}^{i} r_j\right).$$

As a special case, the strife generated by a fuzzy set can be obtained (assuming the set is normal and defined on a discrete universe of discourse) by setting

$$r_1 = F(x_1) \geq r_2 = F(x_2) \geq \cdots \geq r_n = F(x_n) \geq r_{n+1} = 0.$$

The total uncertainty present in a decision problem is the sum of the nonspecificity and the strife (i.e., $N(m) + S(m)$). Often a viable decision is one that minimizes uncertainty. Our formulation therefore allows for decision making;for example, if alternate outputs are fuzzy sets F_i, we select the decision that minimizes $N(m_{F_i}) + S(m_{F_i})$.

For additional work on the theory of evidence and related topics, we refer the reader to [19], [33], and [7].

2.5 Rough Sets and Fuzzy Sets

2.5.1 Introduction

Let X be the universe of discourse. Assume that we have an equivalence relation R on X. Then R is a function from $X \times X$ into $\{0,1\}$ that satisfies

$$\begin{aligned} &R(x,x) = 1 \quad \text{for all} \quad x \in X, \\ &R(x,y) = 1 \quad \text{implies} \quad R(y,x) = 1, \\ &R(x,y) = 1 \quad \text{and} \quad R(y,z) = 1 \quad \text{implies} \quad R(x,z) = 1. \end{aligned}$$

Let $[x]$ denote the class of equivalence of x (i.e., $x = \{y \,|\, R(x,y) = 1\}$). It is well known that distinct classes are disjoint and, moreover, classes of equivalence form a partition of X. Let A be crisp subset of X. We define the lower approximation and the upper approximation of A by

$$\underline{R}(A) = \bigcup_{[x] \subset A} [x] \quad \text{and} \quad \overline{R}(A) = \bigcup_{[x] \cap A \neq \emptyset} [x].$$

A rough set is defined as a representation of a set A by its lower approximation and its upper approximation. Thus, as in fuzzy sets, the boundary is not sharp; it is caught somewhere between the lower and the upper approximation. Generalization of this concept can take two forms: The set A is fuzzy or the set A is crisp but the relation R is fuzzy. If R is a fuzzy relation, it is taken to be a similarity relation; that is,

$$\begin{aligned} &R(x,x) = 1 \quad \text{for all} \quad x \in X, \\ &R(x,y) = 1 \quad \text{implies} \quad R(y,x) = 1, \\ &R(x,z) \geq \max_y \min\{R(x,y), R(y,z)\}. \end{aligned}$$

If R is a similarity relation, we define classes of similarity. If $x \in X$, then $[x]$ denotes the function whose value at y is $R(x,y)$. Of course, it is not true any more that distinct classes are disjoint. In general, if R and A are crisp, then

$$\underline{R}(A) \subset A \subset \overline{R}.$$

If $\underline{R}(A) = A = \overline{R}(A)$, then A is said to be definable. Rough sets are naturally generalized in the context of an information system. By an information system we mean the quadruple $< X, Q, V, \rho >$, where Q is partitioned into two sets C and D (conditions and decisions). The set V is the set of values and ρ is a function from $X \times Q$ into V, where the interpretation of $\rho(x,q)$ is the value of q for x. Table 2.1 illustrates these concepts.

A possible interpretation is the following: There are five patients $x_1, \cdots, x_5$. Two conditions denoted by c_1 and c_2 are observed in these patients. In column c_1, the values 0 and 1 denote the absence or presence of some symptom. In column c_2, L and S denote a large or a small presence of a second symptom.

	C		D
	c_1	c_2	d
x_1	0	L	0
x_2	1	S	1
x_3	1	S	0
x_4	1	L	1
x_5	0	L	0

Table 2.1: patients-symptoms-disease

In the column d, the values 0 and 1 denote the absence or the presence of some disease. The space X is then partitioned into sets of patients having the same symptoms. The partition is $[x_1] = [x_5] = \{x_1, x_5\}$, $[x_2] = [x_3] = \{x_2, x_3\}$, and $[x_4] = \{x_4\}$. Let A denote the set of sick patients. It is clear that A is not definable in terms of symptoms, as x_2 and x_3 have the same symptoms but $x_2 \in A$ and $x_3 \notin A$. How does such a discrepancy come about? Evidently, the physician has used indicators other than c_1 and c_2 to make such a judgment. It may even be the case that the physician is not conscious of how such a judgment was made. Thus, A is not definable in terms of c_1 and c_2 and the boundary of A is not clearly determined. In this case,

$$A = \{x_2, x_4\}, \quad \underline{R}(A) = \{x_4\}, \quad \overline{R}(A) = \{x_2, x_3, x_4\}.$$

It is precisely the size of the set $\overline{R}(A) - \underline{R}(A)$ that reflects the nondefinibility of A in terms of the conditions of Q. In that sense, the lower and the upper approximation of A play a role somewhat analogous to Bel(A) and Pls(A), where Pls(A) – Bel(A) is a reflection of the uncertainty of A. It should be noted that whereas A is not definable in terms of the conditions, $\underline{R}(A)$ and $\overline{R}(A)$ are, in fact,

$$\underline{R}(A) = \{x_i \,|\, c_1 = 1,\, c_2 = L\},$$
$$\overline{R}(A) = \{x_i \,|\, c_1 = 1,\, c_2 = S\} \cap \{x_i \,|\, c_1 = 1,\, c_2 = L\}.$$

2.5.2 Two Operations on Interval Type 2 Sets

The purpose of this subsection is to illustrate by a simple example how one can extract rules from examples using rough sets. The example deals with linking symptoms to diagnosis. In Section 2.2.3 we have seen a number of ways to define $A \to B$, where A and B are fuzzy sets. In this example, for illustrative purposes, we define

$$(A \to B)(x) = (1 - A(x)) \vee B(x).$$

Unfortunately, uncertainty is all too often present in symptoms as well as the diagnosis. Symptoms and diagnosis fail to partition their universes of discourses and overlaps are often present. We now define two operators on fuzzy sets that will be of importance in this subsection. Let A and B be fuzzy sets. We set

$$I(A, B) = \min_x \max\{1 - A(x), B(x)\},$$
$$J(A\#B) = \max_x \min\{A(x), B(x)\}.$$

Note that if A and B are crisp sets, then

$$I(A, B) = \begin{cases} 1 \text{ if } A \subset B \\ 0 \text{ otherwise} \end{cases}$$

and

$$J(A\#B) = \begin{cases} 1 \text{ if } A \cap B \neq \emptyset \\ 0 \text{ otherwise.} \end{cases}$$

Thus, $I(A, B)$ and $J(A, B)$ should intuitively measure how much A is a subset of B and much A intersects B (when A and B are fuzzy). One could also rewrite $I(A, B)$ and $J(A\#B)$ as

$$I(A, B) = \min(A \to B)(x) \quad \text{and} \quad J(A\#B) = \text{Poss}(A, B).$$

It can also be easily checked that

$$I(A, B) = 1 - J(A\#\bar{B}).$$

In a number of situations it might be difficult to exactly determine the membership functions of sets A and B. We thus assume now that A and B are interval type 2 sets; for example,

$$A = \sum [a_i, b_i]/x_i.$$

Thus, the membership of x_i in A is some unknown value in the interval $[a_i, b_i]$. We use interval arithmetic to define the complement of A as

$$1 \ominus A = \sum [1 - b_i, 1 - a_i]/x_i.$$

We also define

$$\tilde{\max}\{[a_i, b_i], [a_j, b_j]\} = [\max(a_i, a_j), \max(b_i, b_j)],$$
$$\tilde{\min}\{[a_i, b_i], [a_j, b_j]\} = [\min(a_i, a_j), \min(b_i, b_j)].$$

Then when A and B are interval type 2 sets, we extend the operators I and J as follows:

$$\tilde{I}(A, B) = \tilde{\min}_x \tilde{\max}(1 \ominus A(x), B(x)),$$
$$\tilde{J}(A\#B) = \tilde{\max}_x \tilde{\min}(A(x), B(x)).$$

Note that for fixed x, $1 \ominus A(x)$ and $B(x)$ are intervals. Thus, $\tilde{\max}(1 \ominus A(x), B(x))$ is an interval that depends on x and, therefore, $\tilde{\min}_x \tilde{\max}(1 \ominus A(x), B(x))$ is also an interval (that does not depend on x). The same observation holds for $\tilde{J}$. Thus, in this more general case, the degree to which A is a subset of B and the degree to which A and B intersect are subintervals of interval $[0, 1]$.

2.5.3 Extracting Rules from Data

We illustrate rule extraction from data by a simple example of two conditions. Condition 1 is the size of a tumor, which could be large or small. Condition 2 is the texture of the tumor, which can be hard or pliable. The diagnosis is that patient is in stage A or stage B. Of course, no sharp boundaries exist between large and small, between hard and pliable, and between stage A and stage B. Intuitively speaking, the operators $\tilde{I}$ and $\tilde{J}$ will generate the lower and the upper approximations to stages A and B. The conditions and the corresponding stages on five patients are shown in the table below.

Patient	Condition 1	Condition 2	Diagnosis
x_1	[0.2,0.4]/L+[0.7,0.9]/S	[0.1,0.3]/H+[0.8,1]/P	$[0.2,0.4]/D_A+[0.5,0.7]/D_B$
x_2	[0.3,0.5]/L+[0.6,0.8]/S	[0.9,1]/H+[0.6,0.9]/P	$[0.6,0.9]/D_A+[0.4,0.6]/D_B$
x_3	[0.6,0.9]/L+[0.2,0.5]/S	[0.5,0.7]/H+[0.5,0.7]/P	$[0.5,0.6]/D_A+[0.8,0.9]/D_B$
x_4	[0.7,0.8]/L+[0.4,0.6]/S	[0.2,0.4]/H+[0.7,0.8]/P	$[0.5,0.8]/D_A+[0.1,0.2]/D_B$
x_5	[0.1,0.4]/L+[0.6,0.7]/S	[0.1,0.2]/H+[0.5,0.9]/P	$[0.3,0.4]/D_A+[0.1,0.2]/D_B$

This above table generates the following interval type 2 sets:

$$
\begin{aligned}
L &= [0.2,0.4]/x_1 + [0.3,0.5]/x_2 + [0.6,0.9]/x_3 + [0.7,0.8]/x_4 + [0.1,0.4]/x_5,\\
S &= [0.7,0.9]/x_1 + [0.6,0.8]/x_2 + [0.2,0.5]/x_3 + [0.4,0.6]/x_4 + [0.6,0.7]/x_5,\\
H &= [0.1,0.3]/x_1 + [0.9,1]/x_2 + [0.5,0.7]/x_3 + [0.2,0.4]/x_4 + [0.1,0.2]/x_5,\\
P &= [0.8,1]/x_1 + [0.6,0.9]/x_2 + [0.5,0.7]/x_3 + [0.7,0.8]/x_4 + [0.5,0.9]/x_5,\\
D_A &= [0.2,0.4]/x_1 + [0.6,0.9]/x_2 + [0.5,0.6]/x_3 + [0.5,0.8]/x_4 + [0.3,0.4]/x_5,\\
D_B &= [0.5,0.7]/x_1 + [0.4,0.6]/x_2 + [0.8,0.9]/x_3 + [0.1,0.2]/x_4 + [0.1,0.2]/x_5.
\end{aligned}
$$

Thus, conditions and diagnoses become interval type 2 sets over the universe of patients. The interpretation is straightforward; for example, x_3 is an example of large tumor L with membership between 0.6 and 0.9. These sets in turn generate rules. For example,

$$
\begin{aligned}
L \cap H &= [0.1,0.3]/x_1 + [0.3,0.5]/x_2 + [0.5,0.7]/x_3 + [0.2,0.4]/x_4\\
&\quad +[0.1,0.2]/x_5,\\
\tilde{\max}&(1 \ominus (L \cap H), D_A) =\\
&[0.7,0.9]/x_1 + [0.6,0.9]/x_2 + [0.5,0.6]/x_3 + [0.6,0.8]/x_4 + [0.8,0.9]/x_5,
\end{aligned}
$$

and, therefore, $\tilde{I}((L \cap H), D_A) = [0.5, 0.6]$. This generates the following certain rule (lower approximation): If the tumor is large and hard, the patient's stage is A with belief between 0.5 and 0.6. Proceeding in this way, a total of eight certain (or belief) rules would be collected (four for D_A and four for D_B). A similar computation would produce $\tilde{J}((L \cap P)\#D_A) = [0.5, 0.8]$ and generate the possible rule: If the tumor is large and pliable, the patient's stage is A with possibility between 0.5 and 0.8. This way a total of 16 rules would be generated. Eight of them are certain and eight of them are possible.

For additional information on rough sets, see [35], and for combinations of rough sets and fuzzy sets see, [3].

2.6 Genetic Algorithms

2.6.1 Introduction

Genetic algorithms imitate natural evolution to solve optimization problems by finding a good solution through a search procedure. The basic construct of a generic algorithm is a chromosome that encodes a possible solution. An initial population of chromosomes is initiated and the population is evolved through three operations: reproduction, crossover, and mutation. The reproduction involves the selection of chromosomes that generate high values for the fitness function. A number of steps need to be taken to run a generic algorithm. The first step is to pick a representation for possible solutions. Typically, potential solutions are represented by a string of numbers and/or characters. Such a string is called a chromosome. For example, a neural net could be represented as a string of weights and biases. Thus, the corresponding chromosome would be a string of numbers. The fitness function measures how good a solution represented by a chromosome is. For neural nets, an appropriate fitness function could be defined through some training set $\{(I_1, d_1), \ldots, (I_m, d_m)\}$, where d_i denotes the desired target when input I_i is presented, and the fitness function might be defined as

$$F(\text{SpecificNet}) = -\sum_{i=1}^{m} (d_i - a_i)^2,$$

where a_i is the activation of the net for input I_i.
Reproduction: Given an initial population, chromosomes are selected with probability proportional to their fitness. The selected chromosomes are copied into a set (known as the mating pool).
Crossover: Pairs of chromosomes in the mating pool are selected randomly and corresponding elements of each are swapped randomly. The diagram in Figure 2.33 shows a commonly used crossover.

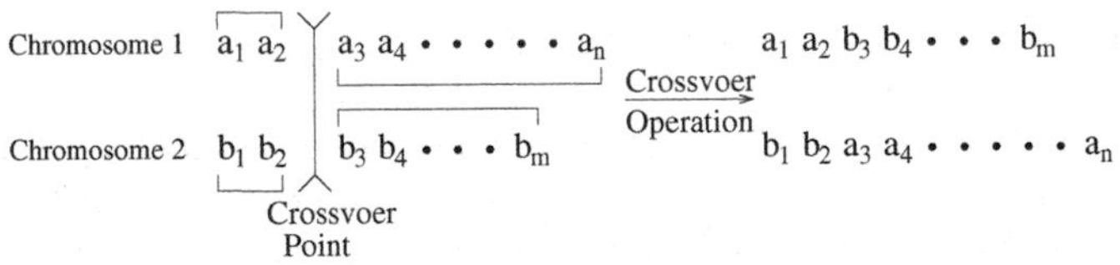

Fig. 2.33

Typically, crossover takes place with some probability and the crossover point is chosen randomly.

em Mutation: After crossover, each element of the string is changed with small probability. Thus, the final result after crossover is performed might be as shown in Figure 2.34.

$a_1\ a_2\ b_3\ a_4\ b_5 \bullet \bullet \bullet\ b_m$

$b_1\ b_2\ a_3\ a_4 \bullet \bullet \bullet \bullet \bullet \bullet\ a_n$

Fig. 2.34

In this example the first string obtained after crossover was subject to a mutation, the b_4 element having changed into a_4 while no mutation took place in the second string.

A typical generic algorithm can be outlined as follows:

1. Create a random initial population of chromosomes (representing potential solutions).

2. Repeat until some termination criterion is met:
   a. Evaluate the fitness of each chromosome.
   b. Reproduce (i.e., select chromosomes with probability proportional to their fitness and enter them in the mating pool).
   c. Crossover pair.
   d. Perform mutation (with small probability).

These new individuals form the new mating pool. Go back to step 2.

The termination criterion could be reached when the best chromosome in the mating pool is fit enough and/or the number of iterations reaches a specified number.

2.6.2 Designing a Fuzzy System

The success of the previously outlined algorithm depends on a number of factors. A key factor is a viable encoding of potential solutions into chromosomes. We illustrate a possible way to encode fuzzy systems (i.e., fuzzy sets of rules). When designing a fuzzy system, it is often not obvious how to partition the input and output spaces, that is, it is not obvious how many rules to use. Typically, the general shape of the membership functions is determined but the exact functions are not, as they depend on parameters. For example, we may decide to use trapezoidal functions, but the widths of their bases could be specified as unknown parameters. A possible chromosome corresponding to a set of fuzzy rules could be constructed as a string consisting of three pairs: the input parameters, the output parameters, and the rule identifiers. To keep the example simple, assume we are dealing with rules of the following form: If x is A, then y is B (i.e., one input, one output). A typical chromosome could be as shown in Figure 2.35.

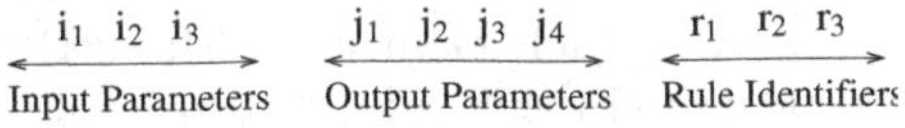

Fig. 2.35

Assume i_1, i_2, and i_3 are the values of the parameters identifying the fuzzy sets A_1, A_2, and A_3, respectively. The values j_1, j_2, j_3, and j_4 identify the fuzzy sets B_1, B_2, B_3, and B_4, respectively. Cells labeled r_1, r_2, and r_3 refer to the rules 1, 2, and 3, respectively. A value $r_i = 0$ means rule i should be taken out.

Let $r_1 = 2$, $r_2 = 0$, and $r_3 = 4$. This chromosome corresponds to the following set of rules:

If x is A_1, then y is B_2.
If x is A_3, then y is B_4.

Assume this chromosome is paired off with the chromosome

$$i'_1\ i'_2\ i'_3\ i'_4\ i'_5\ j'_1\ j'_2\ r'_1\ r'_2\ r'_3\ r'_4\ r'_5,$$

where $i'_1, \cdots, i'_5$ are values of parameters determining the fuzzy sets $A'_1, \cdots, A'_5$, j'_1 and j'_2 determine B'_1 and B'_2, respectively, and $r'_1 = 1$, $r'_2 = 2$, $r'_3 = 1$, $r'_4 = 0$, and $r'_5 = 2$. Then the corresponding system is

If x is A'_1, then y is B'_1.
If x is A'_2, then y is B'_2.
If x is A'_3, then y is B'_1.
If x is A'_5, then y is B'_2.

If we pick a crosspoint between position 2 and 3, after crossover we have the chromosomes

$$i_1\ i_2\ i'_3\ i'_4\ i'_5\ j'_1\ j'_2\ r'_1\ r'_2\ r'_3\ r'_4\ r'_5$$

and

$$i'_1\ i'_2\ i_3\ j_1\ j_2\ j_3\ j_4\ r_1\ r_2\ r_3.$$

Perform a mutation on the first chromosome changing $r'_3 = 1$ to $r'_3 = 0$. The corresponding systems are as following:

If x is A_1, then y is B'_1.
If x is A_2, then y is B'_2.
If x is A'_5, then y is B'_2

and

If x is A'_1, then y is B_2.
If x is A_3, then y is B_4.

The fitness function could be the negative of the error made by the system on a specified training set. After a number of iterations, the best chromosome in the current mating pool would generate the best current fuzzy system.

For more detailed information on genetic algorithms, see [12], and for combining genetic algorithms with fuzzy logic, see [24].

2.7 Conclusion

We presented some of the important methodologies of soft computing. These important tools include neural nets, fuzzy logic, neuro-fuzzy systems, the theory of evidence, rough sets, and genetic algorithms. The main idea that should emerge from this overview is that these methods are not competing but are complementary methods. Adaptive nets use input/output specifications. Such nets form a black box and there is usually no clear indication on how the decisions are made. The advantage of this methodology is that the system learns to adapt as input/output specifications are defined. Fuzzy logic, on the other hand, is a very convenient tool to represent knowledge as a set of rules. Neuro-fuzzy systems combine the advantages of the two previous methods and seamlessly integrate linguistic and numerical information. Two applications were presented: forecasting time series and designing a controller. Although we indicated several ways to construct membership functions, in certain cases such constructions might be difficult to make and some uncertainty regarding membership functions must be taken into consideration. Fuzzy sets of type 2 is a possible approach to this problem.

Fuzzy sets, neural nets, and neuro-fuzzy systems were then compared as possible approaches to designing an autonomous vehicle. The theory of evidence was then presented. This methodology has a built-in capability to use the following type of rules:

If (object$_i$, atribute$_i$, value$_i$), $i = 1, \ldots, n$, then (decision$_1$,...,decision$_k$) (x confirm).

Here, $x \in [0, 1]$. Such a rule generates a mass m. If the antecedent is satisfied, then $m\{\text{decision}_1, \ldots, \text{decision}_k\} = x$. A set of such rules generates a set of masses, and composing these masses yields a mass representing the set of rules. That resulting mass highlights the appropriate decision to consider. Nonspecificity, which is an uncertainty related to how large the set of choices is was discussed in the context of evidence theory. Similarly, the concept of strife, that is the uncertainty related to the conflict between choices, was also cast in the concept of evidence theory. Rough sets were then introduced. As with fuzzy sets, the boundary of a rough set is not a crisp set. There is a lower and an upper approximation. In was then shown how a diagnosis problem in which no sharp boundaries exist between conditions and between diagnoses could generate rules by use of the rough sets methodology. We then have two types of rules: the certain and the possible rules corresponding to lower and upper approximations. Finally, genetic algorithms were introduced. It was shown how a good set of fuzzy rules could be obtained for some specified problem using genetic algorithms.

It is clear that complex problems typically require hybrid methods as the sensible approach. For example, in certain cases it may be convenient to construct the fuzzy part of a neuro-fuzzy system using genetic algorithms and then evolve the neural part using, for example, the swarm method. Future directions might also incorporate the methodology of fuzzy sets of type 2 into such hybrid methods. For hybrid methods and various approaches to uncertainty, we refer the reader to [17] and [20].

Aknowledgment: The authors would like to thank Dr. Youn Sha-Chan who created all figures in this chapter.

References

1. Bandler, W., Kohout, L.J.: On the general theory of relational morphisms. International Journal of General Systems 13, 47–68 (1986)
2. Berenji, H.R., Khedkar, P.: Learning and tuning fuzzy logic controllers through reinforcements. IEEE Transactions on Neural Networks 3(5), 724–740 (1992)
3. Dubois, D., Prade, H.: Putting rough sets and fuzzy sets together. In: R. Slovinski (ed.) Intelligent Decision Support, pp. 203–232. Kluwer Academic Publishers, Norwell, MA (1992)
4. Eberhart, R.C., Shi, Y.: Particle swarm optimization: Developments, applications, and resources. In: Proceedings of the 2001 Congress on Evolutionary Computation CEC2001, pp. 81–86. IEEE Press, Los Alamitos, CA (2001)
5. Grossberg, S.: A neural model of attention, reinforcement and discrimination learning. International Review of Neurobiology 18, 263–327 (1975)
6. Hagan, M., Demuth, H., Beale, M.: Neural Network Design. PWS Publishing Company, Boston, MA (1996)
7. Harmanec, D., Klir, G.J.: Measuring total uncertainty in Dempster-Shafer theory: A novel approach. International Journal of General Systems 22(4), 405–419 (1994)
8. Haykin, S.: Neural Networks: A Comprehensive Foundation. MacMillan, New York (1994)
9. Hecht–Nielsen, R.: Counterpropagation networks. In: M. Caudill, C. Butler (eds.) IEEE First International Conference on Neural Networks (ICNN'87), Vol. II, pp. II-19–32. IEEE, San Diego, CA (1987)
10. Hellendorn H., Thomas C.: Defuzzification in fuzzy controllers. Journal of Intelligent and Fuzzy Systems, 1(2), 109-123 (1993)
11. Hinton, G., Sejnowski, T.: Learning and relearning in Boltzmann machines. In: Rummelhart, D., McClelland, J. (eds.) Parallel Distributed Processing: Explorations in the Microstructure of Cognition. Volume 1: Foundations, pp. 283–335, MIT Press, Cambridge, MA. (1986)
12. Holland, J.H.: Adaptation in Natural and Artificial Systems. University of Michigan Press, Ann Arbor (1975)

13. Jang, J.S.R.: Fuzzy modeling using generalized neural networks and Kalman filter algorithm. In: Proceedings of the Ninth National Conference on Artificial Inteligence (AAAI–91), pp. 762–767 (1991)
14. Jang, J.S.R.: ANFIS: Adaptive-network-based fuzzy inference system. IEEE Transactions on Systems, Man, and Cybernetics 23, 665–684 (1993)
15. Jang, J.S.R., Sun, C.T., Mizutani, E.: Neuro-Fuzzy and Soft Computing. Prentice-Hall, Englewood Cliff, NJ (1997)
16. Kennedy, J., Eberhart, R.C.: Particle swarm optimization. In: Proceedings of the IEEE International Conference on Neural Networks, pp. 1942–1948. IEEE Service Center, Piscataway, NJ (1995)
17. Klir, G.J.: Developments in uncertainty-based information. Advances in Computers 36, 255–332 (1993)
18. Klir, G.J., Yuan, B.: Fuzzy Sets and Fuzzy Logic. Prentice-Hall, Englewood Cliff, NJ (1995)
19. de Korvin, A., Deeba, E., Kleyle, R.: Knowledge acquisition using rough sets when membership values are intervals. Mathematical Modeling and Scientific Computing 1, 470–479 (1993)
20. de Korvin, A., Hashemi, S., Sirisaengtaksin, O.: A body of evidence approach under partially specified environment. Journal of Neural, Parallel and Scientific Computation 13, 91-106 (2005)
21. de Korvin, A., Kleyle, R., Lea, R.: An evidence approach to problem solving when a large number of knowledge systems are available. International Journal of Intelligent Systems 5, 293–306 (1990)
22. de Korvin, A., Modave, F., Kleyle, R.: Paradigms for decision making under increasing levels of uncertainty. International Journal of Pure and Applied Mathematics 21, 419–430 (2005)
23. Kosko, B.: Bidirectional associative memories. IEEE Transactions on Systems, Man, and Cybernetics 18, 49–60 (1988)
24. Lee, M.A., Takagi, H.: Dynamic control of genetic algorithms using fuzzy logic techniques. In: S. Forrest (ed.) Proceedings of the Fifth International Conference on Genetic Algorithms, pp. 76–83. Morgan Kaufmann, San Mateo, CA (1993)
25. Mabuchi, S.: A proposal for defuzzification strategy by the concept of sensitivity analysis. Fuzzy Sets and Systems 55, 1–14 (1993)
26. Mamdami, E.H., Gaines, B.R. (eds.): Fuzzy Reasoning and Its Applications. Academic Press, London (1981)
27. Mamdani, E.H.: Applications of fuzzy logic to approximate reasoning using linguistic systems. IEEE Transactions on Computing 26, 1182–1191 (1977)
28. Mendel, J.: Uncertain Rule-Based Fuzzy Logic Systems: Introduction and New Directions. Prentice-Hall, Upper Saddle River, NJ (2001)
29. Mendel, J.: On the importance of interval sets in type-2 fuzzy logic systems. In: Proceedings Joint 9th IFSA World Congress and 20th NAFIPS International Conference, pp. 1647–1652 (2001)
30. Mendel, J., John, R.: Type-2 fuzzy sets made simple. IEEE Transactions on Fuzzy Systems 10(2), 117–127 (2002)
31. Mizutani, E., Jang, J.S.R.: Coactive neural fuzzy modeling. In: IEEE International Conference on Neural Networks (ICNN'95), Vol. 2, pp. 760–765. IEEE, Perth, Western Australia (1995)
32. Mizutani, E., Jang, J. S. R., Nishio, K., Takagi, H., Auslander, D. M.: Coactive neural networks with adjustable fuzzy membership functions and their appli-

cations. In: International Conference on Fuzzy Logic and Neural Networks, pp. 581–582 (1994)
33. Mrózek, A.: Rough sets and some aspects of expert systems realization. In: Seventh Workshop on Expert Systems and their Applications, pp. 597–611 (1987)
34. Ovchinikov S. V.: Representation of transitive fuzzy relations. In: Skala, Termini, and Trillas (eds.) Aspects of Vagueness, 105–118, Boston, MA (1984)
35. Pawlak, Z.: Rough sets. International Journal of Computer and Information Sciences 11(5), 341–356 (1982)
36. Ruan, D., Kerre, E.E.: Fuzzy implication operators and generalized fuzzy method of cases. Fuzzy Sets Systems 54(1), 23–37 (1993)
37. Saaty, T.L.: Modeling unstructured decision problems: A theory of analytical hierarchies. In: Proceedings of the First International Conference on Mathematical Modeling, University of Missouri- Rolla, Vol. 1, pp. 59–77 (1977)
38. Saaty, T.L.: The Analytic Hierarchy Process. McGraw–Hill, New York (1980)
39. Skapura, D.M.: Building Neural Networks. ACM Press/Addison–Wesley Publishing Co., New York, NY, USA (1995)
40. Takagi, H., Hayashi, I.: NN–driven fuzzy reasoning. International Journal Approximate Reasoning 5, 191–212 (1991)
41. Takagi, H.: Fusion techniques of fuzzy systems and neural networks, and fuzzy systems and genetic algorithms. In: B. Bosacchi, J.C. Bezdek (eds.) Applications of Fuzzy Logic Technology, Proc. of the Society of Photo–Optical Instrumentation Engineers (SPIE) Conference, SPIE Vol. 2061, pp. 402–413 (1993)
42. Tong, R.M.: An annotated bibliography of fuzzy control. In: M. Sugeno (ed.) Industrial Applications of Fuzzy Control, pp. 249–269. Elsevier Science Publishers, Amsterdam (1985)
43. Weber, S.: A general concept of fuzzy connectives, negations and implications based on t-norms and t-conorms. Fuzzy Sets and Systems 11, 115–134 (1983)
44. Yager, R.R.: Approximate reasoning as a basis for rule based expert systems. IEEE Transactions on Systems, Man and Cybernetics 14 (1984)
45. Yager, R.R., Filev, D.P.: On the issue of defuzzification and selection based on a fuzzy set. Fuzzy Sets and Systems 55, 255–273 (1993)

3

Relations Between Interval Computing and Soft Computing

Vladik Kreinovich

Department of Computer Science, University of Texas at El Paso, 500 W. University, El Paso, TX 79968, USA. vladik@utep.edu

This volume is about knowledge processing with interval and soft computing (i.e., about techniques that use both interval and soft computing to process knowledge and about the results of applying these techniques). To better understand these techniques, in Chapter 1 we described fundamentals of interval computing and in Chapter 2 we described the fundamentals of soft computing. Now it is time to explain how these techniques are related and how they can be combined. Some examples of such a relation were already given in Chapter 2 (e.g., interval-valued fuzzy sets). Now it is time to provide a systematic description of this relation. After this chapter, we will be ready to describe how to combine interval and fuzzy techniques and how the resulting combined techniques can be applied to real-life problems.

This chapter starts with a brief reminder of why data processing and knowledge processing are needed in the first place, why interval and fuzzy methods are needed for data and knowledge processing, and which of the possible data and knowledge processing techniques we should use. Then we explain how these reasonable soft computing techniques are naturally related with interval computing. Finally, we explain the need for interval-valued fuzzy techniques - techniques which will be used a lot in our future applications - and how the transition to such techniques is also related to interval computing.

3.1 Why Data Processing and Knowledge Processing Are Needed in the First Place: A Brief Reminder

3.1.1 Classification of Practical Problems

Most practical problems can be crudely classified into three classes:

- We want to *learn* what is happening in the world; in particular, we want to know the numerical values of different quantities (distances, masses, charges, coordinates, etc.).

C. Hu et al. (eds.), *Knowledge Processing with Interval and Soft Computing*, DOI: 10.1007/978-1-84800-326-2_3,

- Based on these values, we would like to *predict* how the state of the world will change over time.
- Finally, we would like to find out what *changes* we need to make in the world so that these changes will lead to the desired results.

A real-life problem often involves solving subproblems of all three types.

This classification is closely related to the well-known classification of practically useful creative activity into engineering and science:

- The tasks of learning the current state of the world and predicting the future state of the world are usually classified as *science*.
- The tasks of finding the appropriate change are usually classified as *engineering*.

For example, measuring the flow of the Rio Grande river at different locations and predicting how this river flow will change over time are problems of science. Finding the best way to change this flow (e.g., by building a levee to protect downtown El Paso) is a problem of engineering.

3.1.2 First Class of Practical Problems: Learning the State of the World

Let us start with the first class of practical problems: the problem of learning the state of the world. As we have mentioned, this means, in particular, that we want to know the numerical values of different quantities y that characterize this state.

Some quantities y we can simply directly measure. For example, when we want to know the current state of a patient in a hospital, we can measure the patient's body temperature, blood pressure, weight, and many other important characteristics. In some situations, we do not even need to measure: We can simply ask an expert, and the expert will provide us with an (approximate) value $\widetilde{y}$ of the quantity y.

However, many other quantities of interest are difficult or even important to measure or estimate directly. Examples of such quantities include the amount of oil in a given well or the distance to a star. Since we cannot directly measure the values of these quantities, the only way to learn some information about them is to measure (or ask an expert to estimate) some other easier-to-measure quantities $x_1, \ldots, x_n$, and then to estimate y based on the measured values $\widetilde{x}_i$ of these auxiliary quantities x_i.

For example, to estimate the amount of oil in a given well, we perform *seismic* experiments: We set up small explosions at some locations and measure the resulting seismic waves at different distances from the location of the explosion. To find the distance to a faraway star, we measure the direction to the star from different locations on Earth (and/or in different seasons) and the coordinates of (and the distances between) the locations of the corresponding telescopes.

To estimate the value of the desired quantity y, we must know the relation between y and the easier-to-measure (or easier-to-estimate) quantities $x_1, \ldots, x_n$. Specifically, we want to use the estimates of x_i to come up with an estimate for y. Thus, the relation between y and x_i must be given in the form of an *algorithm* $f(x_1, \ldots, x_n)$ that transforms the values of x_i into an estimate for y. Once we know this algorithm f and the measured values $\widetilde{x}_i$ of the auxiliary quantities, we can estimate y as $\widetilde{y} = f(\widetilde{x}_1, \ldots, \widetilde{x}_n)$.

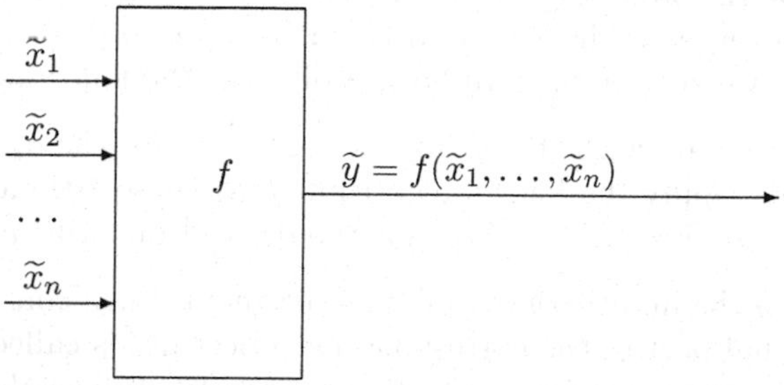

In different practical situations, we have algorithms f of different complexity. For example, to find the distance to star, we can usually have an explicit analytical formula coming from geometry. In this case, f is a simple formula. On the other hand, to find the amount of oil, we must numerically solve a complex partial differential equation. In this case, f is a complex iterative algorithm for solving this equation.

In the case when the values x_i are obtained by measurement, this two-stage process does involve measurement. To distinguish it from *direct* measurements (i.e., measurements which directly measure the values of the desired quantity), the above two-stage process is called an *indirect* measurement.

3.1.3 Second Class of Practical Problems: Predicting the Future State of the World

Once we know the values of the quantities $y_1, \ldots, y_m$ that characterize the current state of the world, we can start predicting the future state of the world (i.e., the future values of these quantities).

To be able to predict the future value z of each of these quantities, we must know how exactly this value z depends on the current values $y_1, \ldots, y_m$. Specifically, we want to use the known estimates $\widetilde{y}_i$ for y_i to come up with an estimate for z. Thus, the relation between z and y_i must be given in the form of an *algorithm* $g(y_1, \ldots, y_m)$ that transforms the values of y_i into an estimate for z. Once we know this algorithm g and the estimates $\widetilde{y}_i$ for the current values of the quantities, we can estimate z as $\widetilde{z} = g(\widetilde{y}_1, \ldots, \widetilde{y}_n)$.

The corresponding algorithm g can sometime be very complicated and time-consuming. This is, for example, how weather is predicted now: Weather prediction requires so many computations that it can only be performed on fast supercomputers.

3.1.4 The General Notion of Data and Knowledge Processing

So far, we have analyzed two classes of practical problems:

- The problem of *learning* the current state of the world (i.e., the problem of indirect measurement); and
- The problem of *predicting* the future state of the world.

From the *practical* viewpoint, these two problems are drastically different. However, as we have seen, from the *computational* viewpoint, these two problems are very similar. In both problems, the following holds:

- We start with the estimates $\widetilde{x}_1, \ldots, \widetilde{x}_n$ for the quantities $x_1, \ldots, x_n$.
- We apply the known algorithm f to these estimates, resulting in an estimate $\widetilde{y} = f(\widetilde{x}_1, \ldots, \widetilde{x}_n)$ for the desired quantity y.

When the inputs come from measurements (i.e., constitute *data*), the computational part of the corresponding procedure is called *data processing.* When the inputs come from experts (i.e., constitute *knowledge*), the computational part of the corresponding procedure is called *knowledge processing.*

3.1.5 Third Class of Practical Problems: How to Change the World

Once we know the current state of the world and we know how to predict the consequences of different decisions (designs, etc.), it is desirable to find the decision (design, etc.) that guarantees the given results. Depending on what we want from this design, we can subdivide all of the problems from this class into two subclasses. In both subclasses, the design must satisfy some constraints. Thus, we are interested in finding a design that satisfies all of these constraints.

- In some practical situations, satisfaction of all of these constraints is all we want. In general, there may be several possible designs that satisfy given constraints. In the problems from the first subclass, we do not have any preferences for one of these designs - any one of them will suffice. Such problems are called the problems of *constraint satisfaction.*
- In other practical situations, we do have a clear preference between different designs x. This preference is usually described in terms of an *objective function* $F(x)$ - a function for which more preferable designs x correspond to larger values of $F(x)$. In such a situation, among all of the designs that satisfy given constraints, we would like to find a design x for which the value $F(x)$ of the given objective function is the largest. Such problems are called *optimization problems.*

3.2 Need for Interval Computations

3.2.1 Need to Take Uncertainty into Account

In the case of data processing, we start with measurement results $\widetilde{x}_1, \ldots, \widetilde{x}_n$. Measurements are never exact. There is a nonzero difference $\Delta x_i \stackrel{\text{def}}{=} \widetilde{x}_i - x_i$ between the (approximate) measurement result $\widetilde{x}_i$ and the (unknown) actual value x_i of the i-th quantity x_i. This difference is called the *measurement error*. The result $\widetilde{y} = f(\widetilde{x}_1, \ldots, \widetilde{x}_n)$ of applying the algorithm f to the measurement results $\widetilde{x}_i$ is, in general, different from the result $y = f(x_1, \ldots, x_n)$ of applying this algorithm to the actual values x_i. Thus, our estimate $\widetilde{y}$ is, in general, different from the actual value y of the desired quantity: $\Delta y \stackrel{\text{def}}{=} \widetilde{y} - y \neq 0$.

In many practical applications, it is important to know not only the desired estimate for the quantity y but also how accurate this estimate is. For example, in geophysical applications, it is not enough to know that the amount of oil in a given oil field is about 100 million tons: It is also important to know how accurate this estimate is. If the amount is 100 ± 10, this means that the estimates are good enough and we should start exploring this oil field. On the other hand, if it is 100 ± 200, this means that it is quite possible that the actual value of the desired quantity y is zero (i.e., that there is no oil at all). In this case, it may be prudent to perform additional measurements before we invest a lot of money into drilling oil wells.

The situation becomes even more critical in medical emergencies: It is not enough to have an estimate of blood pressure or body temperature to make a decision (e.g., whether to perform a surgery), it is important that even with the measurement uncertainty, we are sure about the diagnosis - and if we are not, maybe it is desirable to perform more accurate measurements.

It is therefore desirable to find out the uncertainty Δy caused by the uncertainties Δx_i in the inputs:

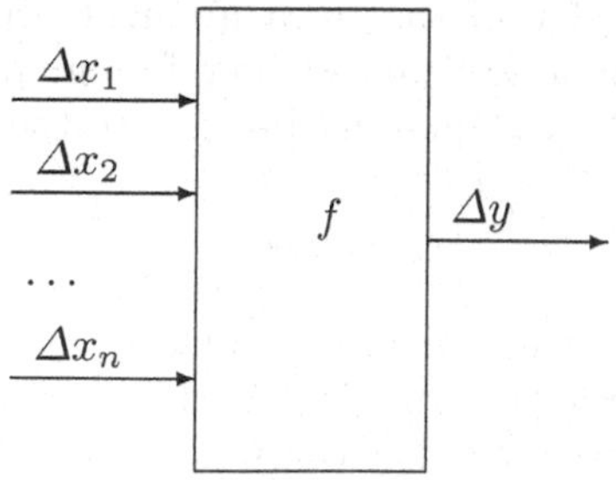

Comment. We assumed that the relation f provides the *exact* relation between the variables $x_1, \ldots, x_n$, and the desired value y. If so, then in the ideal case in which we plug in the actual (unknown) values of x_i into the algorithm f, we get the exact value $y = f(x_1, \ldots, x_n)$ of y.

In many real-life situations, the relation f between x_i and y is only *approximately* known. In this case, even if we know the exact values of x_i, substituting these values into the approximate function f will not provide us with the exact value of y. In such situations, there is even more uncertainty in y:

- First, there is an uncertainty in y caused by the the uncertainty in the inputs.
- Second, there is a *model uncertainty* caused by the fact that the known algorithm f only provides an approximate description of the dependence between the inputs and the output.

A model uncertainty has to be estimated separately and added to the uncertainty caused by the measurement errors.

3.2.2 From Probabilistic to Interval Uncertainty

To estimate the uncertainty Δy caused by the measurement uncertainties Δx_i, we need to have some information about these original uncertainties Δx_i. The whole idea of uncertainty is that we do not know the exact value of x_i (hence, we do not know the exact value of Δx_i). In other words, there are several possible values of Δx_i. Thus, the first thing we would like to know is the *set* of possible values of Δx_i.

We may also know that some of these possible values are more frequent than the others. In other words, we may also have some information about the *probabilities* of different possible values Δx_i.

The manufacturers of a measuring device usually provide us with an upper bound Δ_i for the (absolute value of) possible measurement errors (i.e., with the bound Δ_i for which we are guaranteed that $|\Delta x_i| \leq \Delta_i$).

The need for such a bound comes from the very nature of a measurement process. Indeed, if no such bound is provided, this means that the actual value x_i can be as different from the "measurement result" $\widetilde{x}_i$ as possible. Such a value $\widetilde{x}_i$ is not a measurement; it is a wild guess.

Since the (absolute value of the) measurement error $\Delta x_i = \tilde{x}_i - x_i$ is bounded by the given bound Δ_i, we can therefore guarantee that the actual (unknown) value of the desired quantity belongs to the interval

$$\boldsymbol{x}_i \stackrel{\text{def}}{=} [\widetilde{x}_i - \Delta_i, \widetilde{x}_i + \Delta_i].$$

For example, if the measured value of a quantity is $\widetilde{x}_i = 1.0$ and the upper bound Δ_i on the measurement error is 0.1, this means that the (unknown) actual value of the measured quantity can be anywhere between $1 - 0.1 = 0.9$ and $1 + 0.1 = 1.1$, i.e., that it can take any value from the interval $[0.9, 1.1]$.

In many practical situations, we not only know the interval $[-\Delta_i, \Delta_i]$ of possible values of the measurement error; we also know the probability of different values Δx_i within this interval [8].

In most practical applications, it is assumed that the corresponding measurement errors are normally distributed with 0 mean and known standard deviation. Numerous engineering techniques are known (and widely used) for processing this uncertainty; see, for example, [8].

In practice, we can determine the desired probabilities of different values of Δx_i by comparing the following:

- The result $\widetilde{x}_i$ of measuring a certain quantity with this instrument and
- The result $\widetilde{x}_{i\,\mathrm{st}}$ of measuring the same quantity by a standard (much more accurate) measuring instrument.

Since the standard measuring instrument is much more accurate than the one we use (i.e., $|\widetilde{x}_{i\,\mathrm{st}} - x_i| \ll |\widetilde{x}_i - x_i|$), we can assume that $\widetilde{x}_{i\,\mathrm{st}} = x_i$, and, thus, that the difference $\widetilde{x}_i - \widetilde{x}_{i\,\mathrm{st}}$ between these two measurement results is practically equal to the measurement error $\Delta x_i = \widetilde{x}_i - x_i$. Thus, the empirical distribution of the difference $\widetilde{x}_i - \widetilde{x}_{i\,\mathrm{st}}$ is close to the desired probability distribution for measurement error.

There are two cases, however, when this determination is not done:

- The first is the case of *cutting-edge* measurements (e.g., measurements in fundamental science). When the Hubble telescope detects the light from a distant galaxy, there is no "standard" (much more accurate) telescope floating nearby that we can use to calibrate the Hubble: the Hubble telescope is the best we have.
- The second case is the case of real *industrial* applications (such as measurements on the shop floor). In this case, in principle, every sensor can be thoroughly calibrated, but sensor calibration is so costly - usually costing several orders of magnitude more than the sensor itself - that manufacturers rarely do it (only if it is absolutely necessary).

In both cases, we have no information about the probabilities of Δx_i; the only information we have is the upper bound on the measurement error.

In such cases, after performing a measurement and getting a measurement result $\widetilde{x}_i$, the only information that we have about the actual value x_i of the measured quantity is that it belongs to the interval $\boldsymbol{x}_i = [\widetilde{x}_i - \Delta_i, \widetilde{x}_i + \Delta_i]$. In other words, we do know not the actual value x_i of the i-th quantity. Instead, we know the *interval* $[\widetilde{x}_i - \Delta_i, \widetilde{x}_i + \Delta_i]$ that contains x_i. In this situation, for each i, we know the interval $\boldsymbol{x}_i$ of possible values of x_i, and we need to find the range

$$\boldsymbol{y} \stackrel{\text{def}}{=} \{f(x_1, \ldots, x_n) : x_1 \in \boldsymbol{x}_1, \ldots, x_n \in \boldsymbol{x}_n\}$$

of the given function $f(x_1, \ldots, x_n)$ over all possible tuples $x = (x_1, \ldots, x_n)$, with $x_i \in \boldsymbol{x}_i$. Since the function $f(x_1, \ldots, x_n)$ is usually continuous, this range is also an interval (i.e., $\boldsymbol{y} = [\underline{y}, \overline{y}]$ for some $\underline{y}$ and $\overline{y}$). So, to find this range, it is sufficient to find the endpoints $\underline{y}$ and $\overline{y}$ of this interval.

Let us formulate the corresponding *interval computations* problem in precise terms. We are given the following:

- An integer n;
- n intervals $\boldsymbol{x}_1 = [\underline{x}_1, \overline{x}_1]$, ..., $\boldsymbol{x}_n = [\underline{x}_n, \overline{x}_n]$;

- An algorithm $f(x_1, \ldots, x_n)$ that transforms n real numbers into a real number $y = f(x_1, \ldots, x_n)$.

We need to compute the endpoints $\underline{y}$ and $\overline{y}$ of the interval

$$\boldsymbol{y} = [\underline{y}, \overline{y}] = \{f(x_1, \ldots, x_n) : x_1 \in [\underline{x}_1, \overline{x}_1], \ldots, [\underline{x}_n, \overline{x}_n]\}.$$

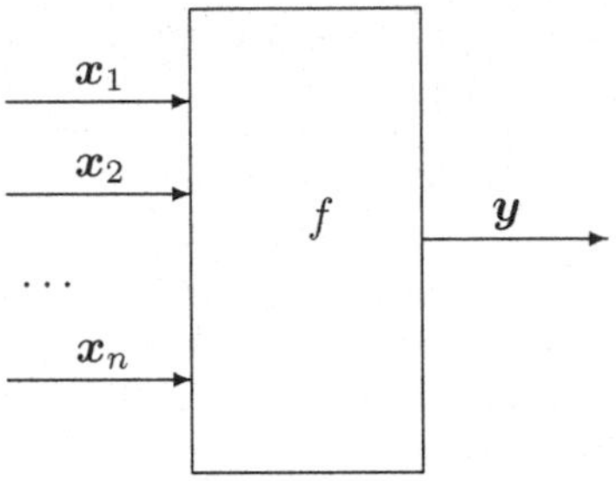

Interval computations are also important for the second class of problems: predicting future values.

3.3 Knowledge Processing and Fuzzy Uncertainty

3.3.1 Need to Process Fuzzy Uncertainty

In many practical situations, we only have expert estimates for the inputs x_i. Sometimes, experts provide guaranteed bounds on the x_i and even the probabilities of different values within these bounds. However, such cases are rare. Usually, the experts' opinions about the uncertainty of their estimates are described by (imprecise, "fuzzy") words from natural language. For example, an expert can say that the value x_i of the i-th quantity is approximately equal to 1.0, with an accuracy most probably of about 0.1. Based on such "fuzzy" information, what can we say about $y = f(x_1, \ldots, x_n)$?

The need to process such "fuzzy" information was first emphasized in the early 1960s by Zadeh, who designed a special technique of *fuzzy logic* for such processing; see, for example, [1, 7].

3.3.2 Processing Fuzzy Uncertainty: Main Idea

Intuitively, a value y is a reasonable value of the desired quantity if $y = f(x_1, \ldots, x_n)$ for some reasonable values x_i (i.e., if for some values $x_1, \ldots, x_n$, x_1 is reasonable, x_2 is reasonable, ..., and $y = f(x_1 \ldots, x_n)$). Thus, to describe to what extent different values of y are reasonable, we must be able to do the following:

- Describe to what extent (to what degree) different values of x_i are reasonable.
- Combine these degrees into the desired degree of belief in reasonability of y.

3.3.3 Degrees of Belief

Let us first introduce the basic concept of degrees of belief. For example, we would like to estimate to what extent the value $x_i = 0.89$ is consistent with the statement "the value x_i of the i-th quantity is approximately equal to 1.0, with an accuracy most probably about 0.1."

In the absence of uncertainty, every statement is either true or false. In the computer, "true" is usually represented as 1 and "false" as 0. It is therefore reasonable to use numbers between 0 and 1 to represent levels of confidence that are intermediate - intermediate between the absolute confidence that a given statement is true and the absolute confidence that a given statement is false.

How do we determine this degree of confidence? Some methods have been described in the previous chapter. For example, we can ask several (N) experts whether $x_i = 0.89$ is consistent with the above statement, and if M of them reply "yes", take the ratio M/N as the desired degree of confidence. If we do not have access to numerous experts, we can simply ask the only available expert to describe his or her degree of confidence by marking a number on a scale from 0 to N (e.g., on a scale from 0 to 5). If an expert marks his or her degree as M, we take the ratio M/N as the desired degree of confidence.

3.3.4 Membership Functions

To formally describe the original expert's statement S about x_i, we need to know, for every real number x_i, the degree $\mu_S(x_i)$ to which this real number is consistent with this statement S.

By using the above procedure, we can determine this value $\mu_S(x_i)$ for every given real number x_i. This procedure includes asking questions of the expert. In practice, we can only ask finitely many questions. Thus, no matter how many questions we ask, by using the above procedure we can only find the values $\mu_S(x_i)$ for finitely many real numbers x_i. To estimate the values $\mu_S(x_i)$ for all other real numbers x_i, we must therefore use interpolation and extrapolation. As we have mentioned in the previous chapter, usually, a piecewise interpolation is used, but sometimes a more sophisticated procedure is applied (e.g., a piecewise quadratic interpolation).

The function $\mu_S(x_i)$ that is obtained by this approximation is called a *membership function.* This function describes, for every real number x_i, the degree $\mu_S(x_i)$ to which this real number is consistent with this statement S.

3.3.5 Need for "And" and "Or" Operations: t-Norms and t-Conorms

As we have mentioned earlier, we are not directly interested in the degree to which a given real number x_i is consistent with the expert's knowledge S_i about the i-th input. We are mainly interested in the degree to which x_1 is consistent with the knowledge about the first input *and* x_2 is consistent with the knowledge about the second input *and* ... *and* x_n is consistent with the knowledge about the n-th input.

In principle, we can determine the degree of belief in such a composite statement by asking an expert, for each possible combination of values $x_1, x_2, \ldots, x_n$, what is the degree to which this combination is consistent with all the available expert knowledge. However, as we have mentioned earlier, even for a single input, we cannot realistically elicit degrees of confidence about too many values. If we consider N possible values of each input, then we would need to elicit the expert's degree of confidence about $N^n \gg N$ possible combinations - which is even less realistic.

Since we cannot directly elicit the expert's degree of confidence in all composite statements, a natural idea is to estimate the degree of confidence in the composite statement based on the degrees of confidence in individual statements, such as "x_i is consistent with the expert's knowledge S_i about the i-th input."

How can we come up with such an estimate? Let us reformulate this estimation problem:

- We know the expert's degree of confidence in statements A_1, A_2, ..., A_n.
- We want to estimate the expert's degree of confidence in a composite statement $A_1 \,\&\, A_2 \,\&\, \ldots \,\&\, A_n$ (i.e., "A_1 and A_2 and ... and A_n").

Since, for example, $A_1 \,\&\, A_2 \,\&\, A_3$ can be represented as $(A_1 \,\&\, A_2) \,\&\, A_3$, it is sufficient to solve this estimation problem for the case of two statements. Once we have a solution for this particular case, we will then be able to solve the general problem as well:

- First, we apply the two-statement solution to the degrees of certainty in A_1 and A_2 and get an estimate for the expert's degree of certainty in $A_1 \,\&\, A_2$.
- Then we apply the same solution to the degrees of certainty in $A_1 \,\&\, A_2$ and A_3 and get an estimate for the expert's degree of certainty in $A_1 \,\&\, A_2 \,\&\, A_3$.
- After that, we apply the same solution to the degrees of certainty in $A_1 \,\&\, A_2 \,\&\, A_3$ and A_4 and get an estimate for the expert's degree of certainty in $A_1 \,\&\, A_2 \,\&\, A_3 \,\&\, A_4$.
- Etc.

Eventually, we will get the degree of confidence in the desired composite statement $A_1 \,\&\, A_2 \,\&\, \ldots \,\&\, A_n$.

Thus, we need a procedure that would transform the degree of belief d_1 in a statement A_1 and the degree of belief d_2 in a statement A_2 into a (reasonable) estimate for a degree of belief in a composite statement $A_1 \,\&\, A_2$. Let us denote the estimate corresponding to given values d_1 and d_2 by $f_\&(d_1, d_2)$. The procedure $f_\&$ that maps degrees of belief d_1 and d_2 in statements A_1 and A_2 into a degree of belief $d = f_\&(d_1, d_2)$ in $A_1 \,\&\, A_2$ is called an *"and" operation*, or, for historical reasons, a *t-norm*.

Similarly, to estimate the degree of belief in a composite statement $A_1 \vee A_2$ ("A_1 or A_2"), we need a procedure $f_\vee$ that maps degrees of belief d_1 and d_2 in statements A_1 and A_2 into a degree of belief $d = f_\vee(d_1, d_2)$ in $A_1 \vee A_2$. Such a procedure is called an *"or"-operation* . Since, in logic, "or" is a kind of dual to "and", an "or" operation can be viewed as a dual to an "and" operation (t-norm). Because of this duality, an "or" operation is also called a *t-conorm.*

3.3.6 Properties of "And" and "Or" Operations

From the intended meaning of the "and" and "or" operations, we can deduce reasonable properties of these operations. For example, intuitively, "A_1 and A_2" means the same as "A_2 and A_1." Thus, it is reasonable to require that our estimate $f_\&(d_1, d_2)$ for the degree of confidence in "A_1 and A_2" should be the same our estimate $f_\&(d_2, d_1)$ for the degree of confidence in "A_2 and A_1." In other words, we must have $f_\&(d_1, d_2) = f_\&(d_2, d_1)$ for all possible values of d_1 and d_2. In mathematical terms, this means that the function $f_\&$ must be *commutative.*

Similarly, "(A_1 and A_2) and A_3" means the same as "A_1 and (A_2 and A_3)" because both mean the same as "A_1 and A_2 and A_3." For each "and" operation $f_\&$, the expression "(A_1 and A_2) and A_3" means that we

- first estimate the degree of belief in "A_1 and A_2" as $f_\&(d_1, d_2)$, and
- then estimate the degree of belief in "(A_1 and A_2) and A_3" as

$$f_\&(f_\&(d_1, d_2), d_3).$$

Similarly, the expression "A_1 and (A_2 and A_3)" means that we

- first estimate the degree of belief in "A_2 and A_3" as $f_\&(d_2, d_3)$ and
- then estimate the degree of belief in "A_1 and (A_2 and A_3)" as

$$f_\&(d_1, f_\&(d_2, d_3)).$$

Since the expressions are equivalent, it is reasonable to require that these estimates coincide (i.e., that $f_\&(f_\&(d_1, d_2), d_3) = f_\&(d_1, f_\&(d_2, d_3))$ for all possible values of d_1, d_2, and d_3). In mathematical terms, this means that the function $f_\&$ must be *associative.*

There are several other reasonable properties of "and" operations. For example, since "A_1 and A_2" implies A_1, our degree of belief in the composite statement "A_1 and A_2" cannot exceed our degree of belief in A_1. Thus, it is reasonable to require that the estimate $f_\&(d_1, d_2)$ for this degree of belief should also not exceed our degree of belief d_1 in the statement A_1. In other words, we should have $f_\&(d_1, d_2) \le d_1$ for all possible values of d_1 and d_2.

If A_1 is absolutely true (i.e., $d_1 = 1$), then, intuitively, the composite statement "A_1 and A_2" has exactly the same truth value as A_2. Thus, it is reasonable to require that $f_\&(1, d_2) = d_2$ for all possible values of d_2.

On the other hand, if A_1 is absolutely false (i.e., $d_1 = 0$), then the composite statement "A_1 and A_2" should also be absolutely false, no matter how much we may believe in A_2. Thus, it is reasonable to require that $f_\&(0, d_2) = 0$ for all possible values of d_2.

Finally, if, due to new evidence, our degree of belief in one of the statements A_1 and A_2 increases, the resulting degree of belief in a composite statement "A_1 and A_2" will either increase or stay the same - but it cannot decrease. Thus, it is it is reasonable to require that the operation $f_\&$ be *monotonic* in the sense that if $d_1 \le d_1'$ and $d_2 \le d_2'$, then $f_\&(d_1, d_2) \le f_\&(d_1', d_2')$.

All of these properties are indeed required of an "and" operation (t-norm). Similarly, it is reasonable to require that an "or" operation (t-conorm) $f_\vee$ should be commutative, associative, monotonic, and satisfy the conditions that $d_1 \le f_\vee(d_1, d_2)$, $f_\vee(1, d_2) = 1$ and $f_\vee(0, d_2) = d_2$ for all possible values of d_1 and d_2.

3.3.7 Simplest "And" and "Or" Operations: Derivation

There exist many different "and" and "or" operations that satisfy the above properties; see, for example, [1, 4, 5, 7]. In some applications such as fuzzy control (see Chapter 2), it is crucial to select appropriate operations because we can use the corresponding additional degrees of freedom to tune the resulting control and make it an even better fit for the corresponding objective function.

However, in knowledge processing, when we are very uncertain about the inputs, it is probably more reasonable to select the simplest "and" and "or" operations that are consistent with the expert knowledge. To select such operations, it makes sense to consider yet another property of "and" and "or": that for every statement A, "A and A" means the same as simply A. Thus, it is reasonable to require that for every statement A with a degree of confidence d, our estimate $f_\&(d, d)$ of the expert's degree of confidence in "A and A" should be the same as the original degree of confidence d in the original statement A. Thus, it is reasonable to require that $f_\&(d, d) = d$ for all possible values of d. In mathematical terms, this means that the function $f_\&$ must be *idempotent*.

Similarly, since "A or A" means the same as simply A, it is reasonable to require that $f_\vee(d, d) = d$ for all possible values of d (i.e., that the function $f_\&$ must also be *idempotent*).

It turns out that this additional requirement leads to a unique "and" operation and a unique "or" operation. Let us first show that the only idempotent "and" operation is $f_\&(d_1, d_2) = \min(d_1, d_2)$. Without loss of generality, let us assume that $d_1 \le d_2$. In this case, the desired equality takes the form $f_\&(d_1, d_2) = d_1$. Since the operation $f_\&$ is idempotent, we have $f_\&(d_1, d_1) = d_1$. Due to $d_1 \le d_2$, monotonicity implies that $f_\&(d_1, d_1) \le f_\&(d_1, d_2)$, hence $d_1 \le f_\&(d_1, d_2)$. On the other hand, for an "and" operation, we always have $f_\&(d_1, d_2) \le d_1$. So, we can conclude that $f_\&(d_1, d_2) = d_1$ (i.e., indeed, $f_\&(d_1, d_2) = \min(d_1, d_2)$).

Let us now prove that the only idempotent "or" operation is $f_\vee(d_1, d_2) = \max(d_1, d_2)$. Without loss of generality, let us again assume that $d_1 \le d_2$. In this case, the desired equality takes the form $f_\vee(d_1, d_2) = d_2$. Since the operation $f_\vee$ is idempotent, we have $f_\vee(d_2, d_2) = d_2$. Due to $d_1 \le d_2$, monotonicity implies that $f_\vee(d_1, d_2) \le f_\vee(d_2, d_2)$, hence $f_\vee(d_1, d_2) \le d_2$. On the other hand, for an "or" operation, we always have $d_2 \le f_\vee(d_1, d_2)$. Thus, we conclude that $f_\vee(d_1, d_2) = d_2$ (i.e., indeed, $f_\vee(d_1, d_2) = \max(d_1, d_2)$).

The operations $f_\&(d_1, d_2) = \min(d_1, d_2)$ and $f_\vee(d_1, d_2) = \max(d_1, d_2)$ were actually the first designed by Zadeh; they are still actively used in various applications of fuzzy techniques; see, for example, [1, 7].

3.3.8 Zadeh's Extension Principle

Let us apply the above simple operations to knowledge processing, or, to be more precise, to processing fuzzy uncertainty. In this situation:

- We know an algorithm $y = f(x_1, \ldots, x_n)$ that relates the value of the desired difficult-to-estimate quantity y with the values of easier-to-estimate auxiliary quantities $x_1, \ldots, x_n$.
- We also have expert knowledge about each of the quantities x_i. For each i, this knowledge is described in terms of the corresponding membership function $\mu_i(x_i)$. For each i and for each value x_i, the value $\mu_i(x_i)$ is the degree of confidence that this value is indeed a possible value of the i-th quantity.

Based on this information, we want to find the membership function $\mu(y)$ that describes, for each real number y, the degree of confidence that this number is a possible value of the desired quantity.

As we have mentioned earlier, y is a possible value of the desired quantity if for some values $x_1, \ldots, x_n$, x_1 is a possible value of the first input quantity, and x_2 is a possible value of the second input quantity, ..., and $y = f(x_1 \ldots, x_n)$. We know that the degree of confidence that x_1 is a possible value of the first input quantity is equal to $\mu_1(x_1)$, that the degree of confidence that x_2 is a possible value of the second input quantity is equal to $\mu_2(x_2)$, and so on. The degree of confidence $d(y, x_1, \ldots, x_n)$ in an equality $y = f(x_1 \ldots, x_n)$ is, of course, equal to 1 if this equality holds and to 0 if this equality does not hold.

We have already agreed to represent "and" as min. Thus, for each combination of values $x_1, \ldots, x_n$, the degree of confidence in a composite statement "x_1 is a possible value of the first input quantity, and x_2 is a possible value of the second input quantity, ..., and $y = f(x_1 \ldots, x_n)$" is equal to

$$\min(\mu_1(x_1), \mu_2(x_2), \ldots, d(y, x_1, \ldots, x_n)).$$

We can simplify this expression if we consider two possible cases: when the equality $y = f(x_1 \ldots, x_n)$ holds and when this equality does not hold.

When the equality $y = f(x_1 \ldots, x_n)$ holds, we get $d(y, x_1, \ldots, x_n) = 1$, and, thus, the above degree of confidence is simply equal to

$$\min(\mu_1(x_1), \mu_2(x_2), \ldots, \mu_n(x_n)).$$

When the equality $y = f(x_1 \ldots, x_n)$ does not hold, we get $d(y, x_1, \ldots, x_n) = 0$, and, thus, the above degree of confidence is simply equal to 0.

We want to combine these degrees of belief into a single degree of confidence that "for some values $x_1, \ldots, x_n$, x_1 is a possible value of the first input quantity, and x_2 is a possible value of the first input quantity, ..., and $y = f(x_1 \ldots, x_n)$." The words "for some values $x_1, \ldots, x_n$" means that the following composite property holds either for one combination of real numbers $x_1, \ldots, x_n$, or for another combination - until we exhaust all (infinitely many) such combinations. We have already agreed to represent "or" as max. Thus, the desired degree of confidence $\mu(y)$ is equal to the maximum of the degrees corresponding to different combinations $x_1, \ldots, x_n$. Since we have infinitely many possible combinations, the maximum is not necessarily attained, so we should, in general, consider supremum instead of maximum:

$$\mu(y) = \sup \min(\mu_1(x_1), \mu_2(x_2), \ldots, d(y, x_1, \ldots, x_n)),$$

where the supremum is taken over all possible combinations.

Since we know that the maximized degree is nonzero only when $y = f(x_1 \ldots, x_n)$, it is sufficient to only take the supremum over such combinations. For such combinations, we can omit the term $d(y, x_1, \ldots, x_n)$ in the maximized expression, so we arrive at the following formula:

$$\mu(y) = \sup\{\min(\mu_1(x_1), \mu_2(x_2), \ldots, \mu_n(x_n)) : y = f(x_1, \ldots, x_n)\}.$$

This formula describes a reasonable way to extend an arbitrary data processing algorithm $f(x_1, \ldots, x_n)$ from real-valued inputs to a more general case of fuzzy inputs. It was first proposed by Zadeh and is thus called *Zadeh's extension principle.*

This is the main formula that describes knowledge processing under fuzzy uncertainty. In the following section, we will show that from the computational viewpoint, the application of this formula can be reduced to interval computations - and indeed, this is how knowledge processing under fuzzy uncertainty is usually done, by using this reduction; see, for example, [1, 3, 7].

3.4 Main Relation Between Interval Computing and Soft Computing: Fuzzy-Related Knowledge Processing Can Be Reduced to Interval Computations

3.4.1 An Alternative Set Representation of a Membership Function: *alpha*-Cuts

To describe the desired relation between fuzzy and interval data processing, we must first reformulate fuzzy techniques in an interval-related form.

In some situations, an expert knows exactly which values of x_i are possible and which are not. In this situation, the expert's knowledge can be naturally represented by describing the *set* of all possible values.

In general, the expert's knowledge is fuzzy:

- We may still have some values about which the expert 100% believes that they are possible.
- we may still have some values about which the expert 100% believes that they are impossible.
- However, in general, the expert is not 100% confident about which values of x_i are possible and which are not.

For example, a geophysicist may be confident that the density x_i of some mineral can take on values ranging from 3.4 to 3.7 g/cm^3 and she may know that values smaller than 3.0 or larger than 4.0 are absolutely impossible, but she is not sure whether values from 3.0 to 3.4 or from 3.7 to 4.0 are indeed realistically possible.

As we have mentioned, the ultimate purpose of the measurements and estimates is to make decisions. In the geophysical example, we have measured the density at a certain depth, and we need to decide the following:

- Whether it is possible that we have the desired mineral - in which case we should undertake more measurements.
- Whether it is not possible that we have the desired mineral - in which case we should not waste our resources on this region and move to more promising regions.

In practice, decisions are made under uncertainty. If we only have a fuzzy expert description of possible values - in terms of the membership function $\mu_S(x_i)$ - which values x_i should we then classify as possible ones and which as impossible?

Under uncertainty, a reasonable idea is to select a threshold $\alpha \in (0, 1]$. In this case,

- all the values x_i for which the expert's degree of confidence is strong enough (i.e., for which $\mu_S(x_i) \geq \alpha$) are classified as possible;
- similarly, all the values x_i for which the expert's degree of confidence is not sufficiently strong (i.e., for which $\mu_S(x_i) < \alpha$) are classified as impossible.

The resulting set of possible elements

$$\boldsymbol{x}_i(\alpha) \stackrel{\text{def}}{=} \{x_i : \mu_S(x_i) \geq \alpha\}$$

is called the *α-cut* of the membership function $\mu_S(x_i)$.

The choice of a threshold α depends on the practical problem. For example, if we are looking for a potentially very valuable mineral deposit, then it makes sense to continue prospecting even when our degree of confidence is not very high. In this case, it makes sense to select a reasonably small threshold α. On the other hand, if the potential benefit is not high and our resources are limited, it makes sense to limit our search to highly promising regions (i.e., to select a reasonably high threshold α).

To adequately describe the expert knowledge irrespective of an application, we therefore need to know the α-cuts corresponding to different thresholds α. Each α-cut $\boldsymbol{x}_i(\alpha)$ describes the set of values that are possible with degree of confidence at least α.

By definition, α-cuts corresponding to different α are *nested*: When $\alpha \leq \alpha'$, then $\mu_S(x_i) \geq \alpha'$ implies $\mu_S(x_i) \geq \alpha$ and, thus,

$$\boldsymbol{x}_i(\alpha') = \{x_i : \mu_S(x_i) \geq \alpha'\} \subseteq \boldsymbol{x}_i(\alpha) = \{x_i : \mu_S(x_i) \geq \alpha\}.$$

Comment. It is worth mentioning that if we know the α-cuts

$$\boldsymbol{x}_i(\alpha) = \{x_i : \mu_S(x_i) \geq \alpha\}$$

corresponding to all possible values $\alpha \in (0, 1]$, then we can uniquely reconstruct the corresponding membership function $\mu_S(x_i)$. The possibility for such a reconstruction follows from the fact that every real number r is equal to the largest largest value α for which $r \geq \alpha$. In particular, for every x_i, the value $\mu_S(x_i)$ is equal to the largest value α for which $\mu_S(x_i) \geq \alpha$. By definition of the α-cut, the inequality $\mu_S(x_i) \geq \alpha$ is equivalent to $x_i \in \boldsymbol{x}_i(\alpha)$. Thus, for every x_i, the value $\mu_S(x_i)$ can be reconstructed as the largest value α for which $x_i \in \boldsymbol{x}_i(\alpha)$.

Thus, we can alternatively view a membership function as a nested family of α-cuts; see, for example, [3].

3.4.2 Fuzzy Numbers and Intervals

In most practical situations, the membership function starts with 0, continuously increases until a certain value, and then continuously decreases to 0. Such membership function describe usual expert's expressions such as "small," "medium," "reasonably high," "approximately equal to a with an error about σ," and so on. Such examples were given in the previous chapter. Since membership functions of this type are actively used in expert estimates of number-valued quantities, they are usually called *fuzzy numbers*.

For a fuzzy number $\mu_i(x_i)$, every α-cut $\boldsymbol{x}_i(\alpha)$ is an interval. Thus, a fuzzy number can be viewed as a nested family of intervals $\boldsymbol{x}_i(\alpha)$ corresponding to different degrees of confidence.

3.4.3 Simplest "And" and "Or" Operations: Reformulation in Terms of Sets and *alpha*-Cuts

The main formulas for fuzzy computations (i.e., for processing fuzzy data) were derived by using the simplest "and" and "or" operations $f_{\&}(d_1, d_2) = \min(d_1, d_2)$ and $f_{\vee}(d_1, d_2) = \max(d_1, d_2)$, respectively. Thus, before we describe how fuzzy computations can be reduced to interval computations, let us first reformulate these "and" and "or"operations in terms of α-cuts.

Specifically, let us assume that we have two properties A and B that are described by the membership functions $\mu_A(x)$ and $\mu_B(x)$ and, correspondingly, by the α-cuts $\boldsymbol{x}_A(\alpha) = \{x : \mu_A(x) \geq \alpha\}$ and $\boldsymbol{x}_B(\alpha) = \{x : \mu_B(x) \geq \alpha\}$. If we use the simplest "and" operation $f_{\&}(d_1, d_2) = \min(d_1, d_2)$, then the composite property $A \,\&\, B$ ("A and B") is described by the membership function $\mu_{A \,\&\, B}(x) = \min(\mu_A(x), \mu_B(x))$. What are the α-cuts

$$\boldsymbol{x}_{A \,\&\, B}(\alpha) = \{x : \mu_{A \,\&\, B}(x) \geq \alpha\}$$

corresponding to this membership function?

The minimum of two real numbers is greater than or equal to α if and only if both of these numbers are greater than or equal to α. Thus, the condition $\mu_{A \,\&\, B}(x) = \min(\mu_A(x), \mu_B(x)) \geq \alpha$ is equivalent to "$\mu_A(x) \geq \alpha$ and $\mu_B(x) \geq \alpha$." Hence, the set $\boldsymbol{x}_{A \,\&\, B}(\alpha)$ of all the values x for which the condition $\mu_{A \,\&\, B}(x) = \min(\mu_A(x), \mu_B(x)) \geq \alpha$ is satisfied can be found simply as the intersection of the set of all x for which $\mu_A(x) \geq \alpha$ and the set of all x for which $\mu_B(x) \geq \alpha$. In other words, for every α, we have

$$\boldsymbol{x}_{A \,\&\, B}(\alpha) = \boldsymbol{x}_A(\alpha) \cap \boldsymbol{x}_B(\alpha).$$

Therefore, to perform the simplest "and" operation $f_{\&}(d_1, d_2) = \min(d_1, d_2)$, we simply take the intersection of the corresponding α-cuts. This is a very natural operation, since, for exactly defined sets and properties, the set of all the elements that satisfy the property $A \,\&\, B$ is equal to the intersection of the set of all elements that satisfy property A and the set of all elements that satisfy property B.

Similarly, for the simplest "or" operation $f_{\vee}(d_1, d_2) = \max(d_1, d_2)$, the composite property $A \vee B$ ("A or B") is described by the membership function $\mu_{A \vee B}(x) = \max(\mu_A(x), \mu_B(x))$. To find the α-cuts

$$\boldsymbol{x}_{A \vee B}(\alpha) = \{x : \mu_{A \vee B}(x) \geq \alpha\}$$

corresponding to this membership function, we can use the fact that the maximum of two real numbers is greater than or equal to α if and only if one of these numbers is greater than or equal to α. Thus, the condition $\mu_{A\vee B}(x) = \max(\mu_A(x), \mu_B(x)) \geq \alpha$ is equivalent to "$\mu_A(x) \geq \alpha$ or $\mu_B(x) \geq \alpha$." Hence, the set $\boldsymbol{x}_{A\vee B}(\alpha)$ of all the values x for which the condition $\mu_{A\vee B}(x) = \max(\mu_A(x), \mu_B(x)) \geq \alpha$ is satisfied can be found simply as the union of the set of all x for which $\mu_A(x) \geq \alpha$ and the set of all x for which $\mu_B(x) \geq \alpha$. In other words, for every α, we have

$$\boldsymbol{x}_{A\vee B}(\alpha) = \boldsymbol{x}_A(\alpha) \cup \boldsymbol{x}_B(\alpha).$$

Therefore, to perform the simplest "or" operation $f_\vee(d_1, d_2) = \max(d_1, d_2)$, we simply take the union of the corresponding α-cuts. This is also a very natural operation, since, for exactly defined sets and properties, the set of all the elements that satisfy the property $A \vee B$ is equal to the union of the set of all elements that satisfy property A and the set of all elements that satisfy property B.

3.4.4 Fuzzy Computations Can Be Reduced to Interval Computations: Derivation

The main problem of fuzzy computation can be described as follows:

- We know an algorithm $y = f(x_1, \ldots, x_n)$ that relates the value of the desired difficult-to-estimate quantity y with the values of easier-to-estimate auxiliary quantities $x_1, \ldots, x_n$.
- We also know, for every i from 1 to n, a membership function $\mu_i(x_i)$ that describes the expert knowledge about the i-th input quantity x_i.

Our objective is to compute the function

$$\mu(y) = \sup\{\min(\mu_1(x_1), \mu_2(x_2), \ldots, \mu_n(x_n)) : y = f(x_1, \ldots, x_n)\}.$$

Let us now describe this relation in terms of α-cuts. This description will constitute the main relation between fuzzy and interval computing. This relation was first discovered and proved in [2]. To describe this result in precise terms, let us first make some mathematics-related remarks.

The function $y = f(x_1, \ldots, x_n)$ describes the relation between physical quantities. In physics, such a relation is usually continuous. Even when we have seemingly discontinuous transitions (e.g., in phase transitions when, say, the density of water changes into a much smaller density of steam), it is not really a discontinuous transition; it is simply a very fast but still continuous one. In view of this observation, we will assume that the function $y = f(x_1, \ldots, x_n)$ is continuous.

We will also assume the membership functions $\mu_i(x_i)$ are continuous. If we had exact knowledge, then continuity would make no sense, since then the corresponding degree of confidence would abruptly go from 1 for possible values to 0 for impossible ones, without ever attaining any intermediate degrees. However, for fuzzy knowledge, continuity makes perfect sense. If there is some degree of confidence that a value x_i is possible, then it makes sense to assume that values close to x_i are possible too - with a similar degree of belief. In practice, as we mentioned earlier in this chapter and in the previous chapter, membership functions are indeed usually continuous.

It is important to mention that for continuous membership functions $\mu_i(x_i)$, α-cuts $\{x_i : \mu_i(x_i) \geq \alpha\}$ are closed sets (i.e., sets that contain all of their limit points).

Finally, we require that for every i and for every $\alpha > 0$, the α-cut is a compact set. For real numbers, since we have already assumed that the α-cuts $\{x_i : \mu_i(x_i) \geq \alpha\}$ are closed sets, it is sufficient to require that these sets are bounded. This is true, for example, if we assume that all of the membership functions correspond to fuzzy numbers; in this case, all α-cuts are intervals.

Suppose that we know the α-cuts $\boldsymbol{x}_i(\alpha)$ corresponding to the inputs and we want to find the α-cuts $\boldsymbol{y}(\alpha)$ corresponding to the output. By definition of an α-cut, $y \in \boldsymbol{y}(\alpha)$ means that $\mu(y) \geq \alpha$, that is, that

$$\sup\{\min(\mu_1(x_1), \mu_2(x_2), \ldots, \mu_n(x_n)) : y = f(x_1, \ldots, x_n)\} \geq \alpha.$$

By definition of the supremum, this means that for every integer $k > 2/\alpha$, there exists a tuple $(x_1^{(k)}, x_2^{(k)}, \ldots, x_n^{(k)})$ for which $y = f(x_1^{(k)}, \ldots, x_n^{(k)})$ and

$$\min(\mu_1(x_1^{(k)}), \mu_2(x_2^{(k)}), \ldots) \geq \alpha - 1/k.$$

The minimum of several numbers is $\geq \alpha - 1/k$ if and only if all of these numbers are $\geq \alpha - 1/k$ (i.e., $\mu_i(x_i^{(k)}) \geq \alpha - 1/k$ for all i). Since $k > 2/\alpha$, we have $1/k < \alpha/2$ and $\alpha - 1/k > \alpha/2$. Thus, for each i and all k, the value $x_i^{(k)}$ belongs to the compact $(\alpha/2)$-cut $\boldsymbol{x}_i(\alpha/2)$. Since the tuples $(x_1^{(k)}, x_2^{(k)}, \cdots, x_n^{(k)})$ belong to the compact set

$$\boldsymbol{x}_1(\alpha/2) \times \boldsymbol{x}_2(\alpha/2) \times \cdots \times \boldsymbol{x}_n(\alpha/2),$$

the sequence of these tuples has a convergent subsequence converging to some tuple $(x_1, x_2, \ldots, x_n)$. Since both f and μ_i are continuous, for this limit tuple we get $y = f(x_1, \ldots, x_n)$ and $\mu_i(x_i) \geq \alpha$. In other words, every element $y \in \boldsymbol{y}(\alpha)$ can be represented as $y = f(x_1, \ldots, x_n)$ for some values $x_i \in \boldsymbol{x}_i(\alpha)$.

Conversely, if $x_i \in \boldsymbol{x}_i(\alpha)$ and, $y = f(x_1, \ldots, x_n)$, then $\mu_i(x_i) \geq \alpha$ and therefore, $\min(\mu_1(x_1), \mu_2(x_2), \ldots, \mu_n(x_n)) \geq \alpha$ and hence

$$\sup\{\min(\mu_1(x_1), \mu_2(x_2), \ldots, \mu_n(x_n)) : y = f(x_1, \ldots, x_n)\} \geq \alpha$$

(i.e., $\mu(y) \geq \alpha$ and $y \in \boldsymbol{y}(\alpha)$).

Thus, the desired α-cut $\boldsymbol{y}(\alpha)$ consists of exactly values $y = f(x_1, \ldots, x_n)$ for $x_i \in \boldsymbol{x}_i(\alpha)$:

$$\boldsymbol{y}(\alpha) = \{f(x_1, \ldots, x_n) : x_1 \in \boldsymbol{x}_1(\alpha), \ldots, x_n \in \boldsymbol{x}_n(\alpha)\}.$$

This is exactly the range that we defined when we described interval computations, so we can rewrite this formula as

$$\boldsymbol{y}(\alpha) = f(\boldsymbol{x}_1(\alpha), \ldots, \boldsymbol{x}_n(\alpha)).$$

In particular, for fuzzy numbers, when all α-cuts $\boldsymbol{x}_i(\alpha)$ are intervals, computing each α-cut $\boldsymbol{y}(\alpha)$ is exactly the problem of interval computations.

3.4.5 Fuzzy Computations Can Be Reduced to Interval Computations: Conclusion

If the inputs $\mu_i(x_i)$ are fuzzy numbers and the function $y = f(x_1, \ldots, x_n)$ is continuous, then for each α, the α-cut $\boldsymbol{y}(\alpha)$ of y is equal to the range of possible values of $f(x_1, \ldots, x_n)$ as x_i ranges over $\boldsymbol{x}_i(\alpha)$ for all i:

$$\boldsymbol{y}(\alpha) = f(\boldsymbol{x}_1(\alpha), \ldots, \boldsymbol{x}_n(\alpha)).$$

Thus, from the computational point of view, the problem of processing data under fuzzy uncertainty can be reduced to several problems of data processing under interval uncertainty - as many problems as there are α-levels. As we have mentioned, this is not just a theoretical observation: This is exactly how fuzzy data processing is usually performed and this is how interval computations techniques are explained in fuzzy textbooks.

3.5 Auxiliary Relation Between Interval Computing and Soft Computing: Interval-Valued Fuzzy Techniques

3.5.1 Intervals Are Necessary to Describe Degrees of Belief

Earlier, we described an idealized situation in which we can describe degrees of belief by exact real numbers. In practice, the situation is more complicated, because experts cannot describe their degrees of belief precisely; see, for example, [6] and references therein.

Indeed, let us start by reviewing the above-described methods of eliciting degrees of belief. If an expert describes his or her degree of belief by selecting, for example, 8 on a scale from 0 to 10, this does not mean that his or her degree of belief is exactly 0.8: if instead, we ask him or her to select on a scale from 0 to 9, then whatever he or she chooses, after dividing it by 9, we will never get 0.8. If an expert chooses a value 8 on a 0 to 10 scale, then the only thing that we know about the expert's degree of belief is that it is closer to 8 than to 7 or to 9 (i.e., that this degree of belief belongs to the *interval* $[0.75, 0.85]$).

Another possible source of interval uncertainty is when we have *several* experts and their estimates differ. If, for example, two equally good experts point to 7 and 8, then, if we are cautious, we would rather describe the resulting degree of belief as the interval $[0.7, 0.8]$ (or, in view of the above remark, as the interval $[0.65, 0.85]$).

If we determine the degree of belief by polling, then the same argument shows that the resulting numbers are not precise; for example, if 8 out of 10 experts voted for A, then we cannot say that the actual degree of belief is exactly 0.8, because if we repeated this procedure with 9 experts, we will never get exactly 0.8. In this case, there are two other sources of uncertainty: First, picking experts is sort of a random procedure, so the result of voting is a statistical estimate that is not precise (just like a statistical frequency estimate of probability). A better description will be to give an *interval* of possible values of $d(A)$.

The polling method of estimating the degree of belief is based on the assumption that an expert can always tell whether he believes in a given statement S or not. Then we take the ratio $d(S) = N(S)/N$ of the number $N(S)$ of experts who believe in S to the total number N of experts as the desired estimate. For $\neg S$, we thus have $N(\neg S) = N - N(S)$, so $d(\neg S) = N(\neg S)/N = 1 - d(S)$. In reality, an expert is often unsure about S. In this case, instead of dividing the experts into two categories: -those who believe in S and those who do not- we must divide them into *three* categories: those who believe that S is true (we will denote their number by $N(S)$); those who believe that S is false (we will denote their number by $N(\neg S)$); and those who do not have the definite opinion about S -there are $N - N(S) - N(\neg S)$ of them. In this situation, one number is not sufficient to describe the experts' degree of belief in S; we need at least two. There are two ways to describe it: We can describe the degree of belief in S as $d(S) = N(S)/N$ and the degree of belief in $\neg S$ as $d(\neg S) = N(\neg S)/N$. These two numbers must satisfy the condition $d(S) + d(\neg S) \leq 1$. This description is known as *intuitionistic fuzzy logic*. (The reason for the word "intuitionistic" is that this logic is close to the original intuitionistic idea that the law of excluded middle is not always true.)

Alternatively, we can describe the degree of belief $d(S)$ in S and the degree of *plausibility* of S estimating as the fraction of experts who do not consider S impossible, i.e., as $pl(S) = 1 - d(\neg S)$, i.e., as an interval $[d(S), pl(S)]$. This representation corresponds to the *Dempster-Shafer formalism* (see Chapter 2).

So, to describe degrees of belief adequately, we must use *intervals* instead of real numbers.

3.5.2 Interval Computations for Processing Interval-Valued Degrees of Belief: General Idea

For an expert system with interval-valued degrees of belief, the following problem arises: Suppose that we have an expert system whose knowledge base consists of statements $S_1, \ldots, S_N$ and we have an algorithm $f(Q, d_1, \ldots, d_N)$ (called *inference engine*) that for any given query Q, transforms the degrees of belief $d(S_1), \ldots, d(S_N)$ in the statements from the knowledge base into a degree of belief $d(Q) = f(Q, d(S_1), \ldots, d(S_N))$ in Q (e. g., if $Q = S_1 \& S_2$, then $f(d_1, \ldots, d_N) = f_\&(d_1, d_2)$). Suppose now that we know only the *intervals* $\mathbf{d}(S_1), \ldots, \mathbf{d}(S_N)$ that contain the desired degree of belief. Then the degree of belief in Q can take any value from the set

$$f(Q, \mathbf{d}(S_1), \ldots, \mathbf{d}(S_N)) = \{f(Q, d_1, \ldots, d_N) \mid d_i \in \mathbf{d}(S_i)\}.$$

Computing such an interval is a typical problem of *interval computations.*

In particular, since the functions $f_\&$ and $f_\vee$ are increasing in both arguments, we have

$$f_\&([\underline{x}, \overline{x}], [\underline{y}, \overline{y}]) = [f_\&(\underline{x}, \underline{y}), f_\&(\overline{x}, \overline{y})]$$

and

$$f_\vee([\underline{x}, \overline{x}], [\underline{y}, \overline{y}]) = [f_\vee(\underline{x}, \underline{y}), f_\vee(\overline{x}, \overline{y})].$$

For example,

$$\min([\underline{x}, \overline{x}], [\underline{y}, \overline{y}]) = [\min(\underline{x}, \underline{y}), \min(\overline{x}, \overline{y})]$$

and

$$\max([\underline{x}, \overline{x}], [\underline{y}, \overline{y}]) = [\max(\underline{x}, \underline{y}), \max(\overline{x}, \overline{y})].$$

In the following chapters, we will give examples of practical applications of interval-valued fuzzy values.

3.6 Conclusion

In this chapter, we have explained intrinsic and useful relations between interval computing and soft computing - specifically, fuzzy data processing. The main relation is that a fuzzy set (membership function) can be viewed as a nested family of intervals - its α-cuts corresponding to different levels of uncertainty α. From the computational viewpoint, fuzzy data processing can be (and usually is) reduced to level-by-level interval computations with the corresponding α-cuts.

Another relation comes from the fact that it is usually difficult to describe experts' degrees of certainty by exact real numbers. A more adequate description of expert's uncertainty is by an interval. Processing interval-valued degrees of uncertainty also requires interval computations.

References

1. Klir, G.J., Yuan, B.: Fuzzy Sets and Fuzzy Logic. Prentice-Hall, Englewood Cliffs, NJ (1995)
2. Nguyen, H.T.: A note on the extension principle for fuzzy sets. Journal Math. Anal. and Appl. 64, 369–380 (1978)

3. Nguyen, H.T., Kreinovich, V.: Nested intervals and sets: concepts, relations to fuzzy sets, and applications. In: R.B. Kearfott, V. Kreinovich (eds.) Applications of Interval Computations, pp. 245–290. Kluwer Academic Publishers Group, Norwell, MA (1996)
4. Nguyen, H.T., Kreinovich, V.: Applications of Continuous Mathematics to Computer Science. Kluwer Academic Publishers Group, Norwell, MA (1997)
5. Nguyen, H.T., Kreinovich, V.: Methodology of fuzzy control: an introduction. In: H.T. Nguyen, M. Sugeno (eds.) Fuzzy Systems: Modeling and Control, pp. 19-62. Kluwer Academic Publishers Group, Norwell, MA (1998)
6. Nguyen, H.T., Kreinovich, V., Zuo, Q.: Interval-valued degrees of belief: Applications of interval computations to expert systems and intelligent control. International Journal of Uncertainty, Fuzziness, and Knowledge-Based Systems 5, 317–358 (1997)
7. Nguyen, H.T., Walker, E.A.: First Course in Fuzzy Logic. CRC Press, Boca Raton, FL (2006)
8. Rabinovich, S.: Measurement Errors and Uncertainties: Theory and Practice. American Institute of Physics, New York (2005)

4

Interval Matrices in Knowledge Discovery

Chenyi Hu[1] and R. Baker Kearfott[2]

[1] Department of Computer Science, University of Central Arkansas, 201 Donaghey Avenue, Conway, AR 72035-0001, USA. chu@uca.edu
[2] Department of Mathematics, University of Louisiana at Lafayette, Box 4-1010, Lafayette, LA 70504-1010, USA. rbk@louisiana.edu

In this chapter, we study the concepts of interval matrices and related strategies for knowledge discovery. Section 4.1 introduces interval matrices. Section 4.2 discusses representations of an interval matrix. Section 4.3 investigates approximate solutions for interval linear systems of equations. Section 4.4 examines interval versions of singular value decompositions (SVD) and principal component analysis (PCA). Section 4.5 presents a case study, and we draw conclusions in Section 4.6.

4.1 Interval Matrices

Modern technologies collect massive datasets from observations, experiments, and scientific simulation. To obtain knowledge and information, we need to process the collected data that are mostly stored in digital form as points. Numerous computer software packages have been developed to assist. For example, electronic spreadsheets store and process datasets as tables (matrices). In contrast to traditional point presentation in data arrangement and processing, we investigate interval matrix representation of datasets in this chapter.

4.1.1 Why and What Is an Interval Matrix?

In real-world applications, people often care more about qualitative properties than insignificant quantitative differences. Reflected in daily language, most words are qualitative rather than quantitative. Quantitative measurements describe property characteristics more precisely; however, in many cases, relatively small quantitative differences are not qualitatively significant. People have already noticed that precisely matching point data in knowledge processing can be either unnecessary or even misleading. Therefore, organizing data qualitatively as intervals is a reasonable alternative. For example, normal ranges of test results on human physiology are commonly described as ranges

C. Hu et al. (eds.), *Knowledge Processing with Interval and Soft Computing*,
DOI: 10.1007/978-1-84800-326-2_4, © Springer-Verlag London Limited 2008

(intervals) rather than specific point values. By grouping point data into intervals, one can omit insignificant quantitative differences and focus more on qualitative study.

On the other hand, the real world is dynamic and full of uncertainties. Data collected, especially for nonintegral attributes, inevitably contain measurement errors and random noise. Using intervals to represent data, one automatically takes variability and uncertainty into consideration. Storing data attributes as intervals, we obtain an interval-valued table (matrix).

Definition 1. *An interval matrix* $\boldsymbol{A} = \{\boldsymbol{a}_{ij}\}_{m\times n}$ *is an* $m \times n$ *matrix whose entries,* $\boldsymbol{a}_{ij}$*, are intervals.*

For example, $\boldsymbol{A} = \begin{pmatrix} [0, 1] & [-2, 0] & [-4, -2] \\ [-8, -5] & [0, 0] & [20, 25] \end{pmatrix}$ is a 2×3 interval matrix.

Interval matrices have different features from that of traditional point matrices. Therefore, they can be used to model many discrete and even continuous applications in knowledge processing. The following are few of them that will studied in the rest of this book.

- *Rule-based decision making:* In artificial intelligence, knowledge-based agents are designed and implemented to observe environments and to reason about possible courses of actions. In such systems, decisions are usually made by matching input data (relevance of each environment feature) with a certain set of rules. Assume that an environment e contains m features (i.e., $e = (e_1, e_2, \cdots, e_m)^T$), and n possible different decisions, $d_1, d_2, \ldots, d_n$, that could be selected based on the presence of the environment features. Then, a knowledge-based agent may select a specific decision according to an $m \times n$ rule matrix by matching the input e with column vectors of the rule matrix. If the observation vector matches the j-th column of the matrix, then the decision d_j is selected.
  It is certainly impractical to have a rule matrix that contains all possible observations, because there can be an infinite number of different observations but only a finite number of possible decisions. Moreover, observed data can be interval-valued if one takes possible variation and measurement error into consideration. In fact, decisions are often selected based on ranges of parameter values rather than on points. Hence, an interval-valued rule matrix is more appropriate. We study interval-valued rule matrices for decision making in Chapter 6 of this book.
- *Interval-valued matrix games:* Game theory [9] has been widely used in knowledge processing and decision making-systems. The simplest game is a zero-sum one involving only two players. An $m \times n$ matrix $G = \{g_{ij}\}_{m\times n}$ is used to model such a two-person zero-sum game. If a row player R uses his i-th strategy (row) while a column player C selects her j-th choice (column), then R wins (and subsequently C loses) the amount of g_{ij}. The objective of R is to maximize his gain while C tries to minimize her loss.

Theories and algorithms for solving such matrix games have been well established.
However, due to certain forms of uncertainty in real-world applications, outcomes of a matrix game may not be a fixed number, even though the players do not change their strategies. Because the payoffs may vary within a range for fixed strategies, we can use an interval-valued matrix to model such uncertainty. Decision making with interval-valued game matrices is discussed in Chapter 7 of this book.

- *Interval-weighted graphs and networks*: Weighted graphs have broad applications in various areas. In the literature, weights in a graph have been constants. However, in real-world applications, due to some kinds of uncertainties, weights associated with edges can vary within ranges. For instance, traveling time between A and B is usually not exactly 1 hour but between 55 and 67 minutes. To better model the variability of weights in a graph, we should represent weights as intervals instead of constants. Using the adjacency matrix to represent an interval-weighted graph, we obtain an interval-valued matrix. Optimization algorithms with interval-weighted graphs are discussed in Chapter 8 of this book.

In this chapter, we mainly discuss linear-algebra-related topics. They include ways to obtain and represent an interval matrix, different solutions of an interval linear system, and singular value decomposition of an interval matrix.

4.1.2 Obtaining an Interval Matrix

Here let us discuss possible approaches to gathering interval valued data and constructing interval matrices.

As we know, data attribute values often vary from time to time. Also, related knowledge may be valid only for an associated time period. When we process knowledge for a specified time period, the minimum and maximum of a data attribute naturally forms an interval. For example, the NASDAQ daily index values during the week of May 12 to 16, 2008 are intervals:

```
12-May-08: [2,446.36, 2,490.22]
13-May-08: [2,472.58, 2,498.07]
14-May-08: [2,493.58, 2,528.40]
15-May-08: [2,492.95, 2,535.19]
16-May-08: [2,504.18, 2,537.41]
```

As reported in Section 5.5 of Chapter 5 in this book, we apply this min-max approach in financial market case studies. Using the interval-valued matrices, we obtain astonishing computational results. Nevertheless, there are numerous ways other than the min-max approach to obtain interval-valued data and matrices.

Observed data can be associated with measurement error bounds. By adding to or subtracting from the observed value with its estimated error, one can form an interval. For instance, if the length of a rod is measured as 35 mm with absolute error ±0.1 mm, then the length of the rod is within the interval [34.9, 35.1].

Also, statistical confidence intervals can be obtained according to the mean and variance. Fuzzy mean-value intervals can be produced with LR-type fuzzy random variables.

The best way of forming an interval-valued matrix should depend on the application under consideration and the objectives of the study. In other words, we do not think there is a single perfect method that fits all applications.

4.2 Interval Matrix Endpoints and Midpoint-Width Representations

In dealing with variability and uncertainty, an interval matrix $\boldsymbol{A} = \{\boldsymbol{a}_{ij}\}_{m\times n}$ can be a useful model in knowledge processing. However, most existing algorithms in linear algebra for a point-valued matrix cannot be applied directly to interval matrices. For example, since the result of subtracting a nontrivial interval from itself is nonzero,[3] naively applying Gaussian elimination will seldom give useful bounds on the solution set to an interval linear system of equations; however, careful reconsideration of the algorithm may lead to good results; see [2].

To be able to apply existing point matrix algorithms to an interval matrix, it is necessary to reexamine and to extend the existing concepts, theories, and algorithms for processing interval matrices. To do so, we first define interval matrix endpoints and midpoint-width representations as follows.

Definition 2. *Let $\boldsymbol{A} = \{\boldsymbol{a}_{ij}\}_{m\times n}$ be an $m \times n$ interval matrix. The left endpoint matrix of $\boldsymbol{A}$ is the matrix $A_L = \{\underline{a}_{ij}\}_{m\times n}$ and the right endpoint matrix of $\boldsymbol{A}$ is the matrix $A_R = \{\overline{a}_{ij}\}_{m\times n}$. We call A_L and A_R the endpoint representation of the interval matrix $\boldsymbol{A}$.*

By the definition, A_L and A_R are actually the lower and upper bounds of $\boldsymbol{A}$. They are often denoted as $\underline{A}$ and $\overline{A}$ in the literature.

Definition 3. *Let $\boldsymbol{A} = \{\boldsymbol{a}_{ij}\}_{m\times n}$ be an $m \times n$ interval matrix. The midpoint matrix of $\boldsymbol{A}$, A_M, is the matrix $\{m(\boldsymbol{a}_{ij})\}_{m\times n}$, where $m(\boldsymbol{a}_{ij}) = \frac{\underline{a}+\overline{a}}{2}$ and the width matrix of $\boldsymbol{A}$, A_W, is the matrix $\{w(\boldsymbol{a}_{ij})\}_{m\times n}$, where $w(\boldsymbol{a}_{ij}) = (\overline{a}_{ij} - \underline{a}_{ij})$. We call that A_M and A_W the midpoint-width representation of the interval matrix $\boldsymbol{A}$.*

The relationship between the endpoints and midpoint-width representations of an interval matrix $\boldsymbol{A}$ is described by the following proposition.

[3] For example, $[1, 2] - [1, 2] = [-1, 1] \neq 0$.

Proposition 1. *Let* $\boldsymbol{A} = \{\boldsymbol{a}_{ij}\}_{m\times n}$ *be an interval matrix. Then its left endpoint matrix is* $A_L = A_M - A_W/2$ *and its right endpoint matrix is* $A_R = A_M + A_W/2$.

Proof: The i-th row j-th column element of A_L is $\underline{a}_{ij}$. The i-th row j-th column element of $A_M - A_W/2$ is $(\underline{a}_{ij} + \overline{a}_{ij})/2 - (\overline{a}_{ij} - \underline{a}_{ij})/2 = \underline{a}_{ij}$. Hence, $A_L = A_M - A_W/2$. Similarly, we can prove $A_R = A_M + A_W/2$.

Corollary 1. *Given any two matrices among* A_L, A_R, A_M, *and* A_W, *provided that entries of* A_W *are non-negative, one can find the other two.*

The endpoints, midpoint, and width of an interval matrix $\boldsymbol{A}$, A_L, A_R, A_M, and A_W are point matrices. More importantly, they are meaningful in terms of knowledge discovery. Whereas A_L and A_R represent the lower and upper bounds of $\boldsymbol{A}$, A_M is the center, or the arithmetic average, of $\boldsymbol{A}$ and A_W reflects the range of variations of $\boldsymbol{A}$. Hence, we can apply existing algorithms for a point matrix to study the properties of an interval matrix.

Other than using the midpoint matrix of an interval matrix, one may use a "center of gravity" for the interval matrix. According to fuzzy logic, the center of gravity may be different from the midpoint of an interval matrix. However, representing an interval matrix with its center of gravity and modified width matrices is beyond the scope of this chapter. We now conclude this section with an example involving interval matrix endpoints and midpoint-width representations.

Example 1. Find the A_R, A_M, and $\boldsymbol{A}$, provided

$$A_L = \begin{pmatrix} 0 & 6 & -2 & -4 \\ 5 & 2 & 1 & 3 \\ -8 & -1 & 0 & 20 \end{pmatrix},$$

$$A_W = \begin{pmatrix} 1 & 1 & 2 & 2 \\ 1 & 5 & 2 & 0 \\ 3 & 1 & 0 & 5 \end{pmatrix}.$$

Solution: Since $A_L = A_M - A_W/2$, we have

$$A_M = A_L + A_W/2 = \begin{pmatrix} 0.5 & 6.5 & -1 & -3 \\ 5.5 & 4.5 & 2 & 3 \\ -6.5 & -0.5 & 0 & 22.5 \end{pmatrix}.$$

Hence,

$$A_R = A_M + A_W/2 = \begin{pmatrix} 1 & 7 & 0 & -2 \\ 6 & 7 & 3 & 3 \\ -5 & 0 & 0 & 25 \end{pmatrix}.$$

Therefore,

$$\boldsymbol{A} = [A_L,\ A_R] = \begin{pmatrix} [0,\ 1] & [6,\ 7] & [-2, 0] & [-4, -2] \\ [5,\ 6] & [2,\ 7] & [1,\ 3] & [3,\ 3] \\ [-8, -5] & [-1,\ 0] & [0,\ 0] & [20,\ 25] \end{pmatrix}.$$

4.3 Approximated Solutions for Interval Linear Systems of Equations

Finding numerical solutions to linear systems of equations has been a focal point of study in computational linear algebra. Mature computational algorithms, such as Gaussian elimination with scaled partial pivoting, and solid error analysis methods are shining results. Although such results often provide guidelines, it is not trivial to apply them directly to solve interval linear systems of equations.

In solving interval systems, we need to keep in mind the objectives of the computation. In addition, we should take the properties of interval arithmetic into consideration to determine the practicality of computing. Along these lines, we now consider the concept of a solution set for an interval linear system of equations.

To highlight differences between real linear systems of equations and interval systems of equations, consider the following example.

Example 2. (Exact satisfaction of linear interval equations)

1. Consider the linear interval equation $[2,3]\boldsymbol{x} = [4,6]$. The trivial interval $\boldsymbol{x} = [2,2]$ satisfies the equation perfectly, in the sense that the left interval is equal to the right interval.
2. However, with a slight change on the above equation, there does not exist an interval $\boldsymbol{x}$ such that $[2,3]\boldsymbol{x} = [4,5]$. (If there were such an interval $\boldsymbol{x}$, then the lower bound of $\boldsymbol{x}$ must be positive, since the product would otherwise be nonpositive. Hence, $\boldsymbol{x}$ would be a positive interval and $[2,3][\underline{x},\overline{x}] = [2\underline{x}, 3\overline{x}] = [4,5]$, whence $\underline{x} = 4/2 = 2$ and $\overline{x} = 5/3 < \underline{x}$, a contradiction.)

Thus, the very definition of a solution set needs to be considered if we are to use interval computations meaningfully.

We now assume that $\boldsymbol{A}$ is an $n \times n$ interval matrix and $\boldsymbol{b}$ is an interval n-vector . There are various definitions of a solution set to $\boldsymbol{A}x = \boldsymbol{b}$ in the interval computing literature. The most common one, often called simply the "solution set" to an interval system of equations, is the *united solution set*:

$$\{x \in \mathbb{R}^n \mid \exists A \in \boldsymbol{A} \text{ and } \exists b \in \boldsymbol{b} \text{ such that } Ax = b\};$$

that is, the united solution set is the set of all possible solutions as the matrix ranges across the interval matrix $\boldsymbol{A}$ and the right-hand-side vector ranges over the interval vector $\boldsymbol{b}$. However, this mathematical definition does not directly imply a practical algorithm to identify the solution set computationally. In fact, the solution set for an interval linear system is usually no-convex, see [10]. Furthermore, finding the solution set exactly is NP-complete; see, for example, [7]. Thus, existing interval linear solvers, such as the Krawczyk solver, preconditioned interval Gaussian elimination, or the preconditioned interval Gauss–Seidel method, find an interval vector $\boldsymbol{x}$ that encloses the solution set rather than the solution set itself. Hence, not only is $\boldsymbol{Ax} \supseteq \boldsymbol{b}$, but, since the solution set is approximated by an "outer enclosure" $\boldsymbol{x}$, there are usually some $x \in \boldsymbol{x}$ such that there are no $A \in \boldsymbol{A}$ and $b \in \boldsymbol{b}$ satisfying $Ax = b$. However, such false solutions can be very misleading in knowledge discovery.

Alternate definitions of solution sets in the interval literature, stemming largely from work of Shary, appear along with references to earlier work in [3]. To avoid inclusion of false solutions, an appropriate solution set in knowledge discovery is the *tolerable solution set*, defined as follows.

Definition 4. *(Shary et al.) The tolerable solution set to the interval linear system of equations* $\boldsymbol{A}x = \boldsymbol{b}$ *is the set*

$$\{x \in \mathbb{R}^n \quad | \quad \textit{for every } A \in \boldsymbol{A} \textit{ there exists a } b \in \boldsymbol{b} \textit{ with } Ax = b\}.$$

In knowledge reliability, one usually prefers to find an interval vector $\boldsymbol{x}$ such that $\boldsymbol{Ax} \subseteq \boldsymbol{b}$, to avoid false point solutions. In the terminology of the interval computations literature, an *inner approximation* to the tolerable solution set is sought. In the remainder of this chapter, we will refer to such inner approximations to the tolerable solution sets as *inner approximated solutions.* We will also refer to interval vectors $\boldsymbol{x}$ that somehow approximate some kind of solution set to $\boldsymbol{A}x = \boldsymbol{b}$, but for which possibly either $\boldsymbol{Ax} \subset b$ or $\boldsymbol{Ax} \supset \boldsymbol{b}$, simply as *approximated solutions.*

Appropriate questions that need to be answered are the following:

1. For an interval matrix $\boldsymbol{A}_{n\times n}$ and an interval n-vector $\boldsymbol{b}$, is there an interval n-vector $\boldsymbol{x}$ such that $\boldsymbol{Ax} \subseteq \boldsymbol{b}$?
2. How does one determine if one approximated solution is better than another?

The answer for the first question can be either positive or negative depending on $\boldsymbol{A}$ and $\boldsymbol{b}$. The following proposition provides necessary and sufficient conditions, for the one-dimensional case, for existence of an inner approximated solution.

Proposition 2. *Let* $\boldsymbol{a} > 0$ *and* $\boldsymbol{b} \geq 0$ *be two intervals. There exists an interval* $\boldsymbol{x} \geq 0$ *such that* $\boldsymbol{ax} \subseteq \boldsymbol{b}$ *if and only if* $\overline{a}\underline{b} \leq \underline{a}\overline{b}$.

Proof. If such an interval $\boldsymbol{x}$ exists, then $\boldsymbol{x} \geq 0$; this is because $\underline{x} < 0$ and $\boldsymbol{a} > 0$ imply that the lower bound of $\boldsymbol{ax}$ is less than zero, so $\underline{b}$ would also need to be less than zero. Therefore, $\underline{b} \leq \underline{ax} \leq \overline{ax} \leq \overline{b}$. Hence, $\underline{b}/\underline{a} \leq \underline{x}$ and $\overline{x} \leq \overline{b}/\overline{a}$. From $\underline{x} \leq \overline{x}$, we have $\underline{b}/\underline{a} \leq \overline{b}/\overline{a}$ (i. e., $\overline{a}\underline{b} \leq \underline{a}\overline{b}$).

For the converse, assume there is an interval $\boldsymbol{x} \geq 0$ that satisfies $\boldsymbol{ax} \subseteq \boldsymbol{b}$. Then $\underline{b}/\underline{a} \leq \underline{x}$ and $\overline{x} \leq \overline{b}/\overline{a}$. If we also assume $\overline{a}\underline{b} > \underline{a}\overline{b}$, we would then have $\underline{b}/\underline{a} > \overline{b}/\overline{a}$, which in turn would imply $\underline{x} > \overline{x}$, a contradiction. Therefore, $\boldsymbol{ax} \subseteq \boldsymbol{b}$.

Applying the proposition, we can easily verify that the equation $[2,\ 3]x = [4,\ 5]$ does not have an inner approximated solution, since $\overline{a}\underline{b} = 15 > 10 = \underline{a}\overline{b}$.

Corollary 2. *Let* $\boldsymbol{a} > 0$ *and* $\boldsymbol{b} \geq 0$ *be intervals. If* $\overline{a}\underline{b} = \underline{a}\overline{b}$*, then there exists an interval* $\boldsymbol{x}$ *that satisfies* $\boldsymbol{ax} = \boldsymbol{b}$*. Furthermore,* $\boldsymbol{x}$ *is degenerate and*

$$\underline{x} = \overline{x} = \frac{\underline{b}}{\underline{a}} = \frac{\overline{b}}{\overline{a}}.$$

Proof. If $\overline{a}\underline{b} = \underline{a}\overline{b}$, from Proposition 2, there is an interval $\boldsymbol{x} \geq 0$ such that $\boldsymbol{ax} \subseteq \boldsymbol{b}$. Therefore, $\underline{b} \leq \underline{ax}$ and $\overline{ax} \leq \overline{b}$. Since $\overline{a}\underline{b} = \underline{a}\overline{b}$, we have $\underline{b}/\underline{a} = \overline{b}/\overline{a}$. Since $\underline{b}/\underline{a} \leq \underline{x} \leq \overline{x} \leq \overline{b}/\overline{a}$, we have $\underline{x} = \overline{x} = \underline{b}/\underline{a} = \overline{b}/\overline{a}$.

From the corollary, the equation $[2,\ 3]\boldsymbol{x} = [4,\ 6]$ has a degenerate solution $x = 2$. However, $\overline{a}\underline{b} = \underline{a}\overline{b}$ is not a necessary condition for the existence of $\boldsymbol{x}$ that satisfies $\boldsymbol{ax} \subseteq \boldsymbol{b}$. For example, the interval $\boldsymbol{x} = [2, 3]$ satisfies $[1, 2]\boldsymbol{x} = [2, 6]$, but $\overline{a}\underline{b} = 4 < 6 = \underline{a}\overline{b}$.

As mentioned, we prefer an inner approximation to the tolerable solution set rather than an enclosure of the united solution set when dealing with knowledge reliability. It is a valid open mathematical question to find necessary and sufficient conditions for the existence of an inner approximation for a general $n \times n$ interval linear system of equations. However, in the practice of knowledge processing, one may select a satisfactory approximated solution. We need a notion to compare approximated solutions. To do so, we use the ratio of the volumes of $\boldsymbol{b}$ and $\boldsymbol{Ax}$ as a quality measurement of an approximated solution. The *volume* of an interval vector $\boldsymbol{y} = (\boldsymbol{y}_1, \boldsymbol{y}_2, \ldots, \boldsymbol{y}_n)$ is denoted by

$$v(\boldsymbol{y}) = \prod_{1 \leq i \leq n} (\overline{y}_i - \underline{y}_i).$$

It is obvious that the volume of an interval is the same as the width of that interval.

Definition 5. *Let* $\boldsymbol{x}$ *be an approximated solution of* $\boldsymbol{A}x = \boldsymbol{b}$*. The ratio of the estimation is defined to be*

$$\begin{cases} 0 & \text{if } \boldsymbol{Ax} \cap \boldsymbol{b} = \emptyset; \\ 1 & \text{if } \boldsymbol{Ax} = \boldsymbol{b}; \\ \dfrac{v(\boldsymbol{b})}{v(\boldsymbol{Ax})} & \text{if } \boldsymbol{Ax} \supset \boldsymbol{b}; \\ \dfrac{v(\boldsymbol{Ax})}{v(\boldsymbol{b})} & \text{if } \boldsymbol{Ax} \subset \boldsymbol{b}; \\ \dfrac{v(\boldsymbol{Ax} \cap \boldsymbol{b})}{v(\boldsymbol{Ax} \cup \boldsymbol{b})} & \text{otherwise.} \end{cases}$$

Although the ratio of estimation is a good indicator, it should not be the only consideration. Since an inner approximated solution relates to information reliability, it should be considered first in knowledge processing.

Example 3. Find approximated solutions for the interval equation $[2, 3]\boldsymbol{x} = [4, 5]$ and then compare their ratios of estimation.

Solution: As we discussed in Example 1, there does not exist an interval $\boldsymbol{x}$ such that $[2,3]\boldsymbol{x} = [4,5]$ exactly, nor does there exist an inner approximation. To obtain an interval $\boldsymbol{x}$, we apply the following approaches:

- First, we apply the naive approach of dividing the interval $[4,5]$ by the coefficient $[2,3]$ with interval arithmetic. We obtain $\boldsymbol{x} \approx [4,5]/[2,3] = [4/3, 5/2]$ and $[2,3][4/3,5/2] = [8/3, 15/2] \supset [4,5]$. The ratio of the estimation is
$$\frac{5-4}{15/2 - 8/3} = 6/29 \approx 20.7\%.$$
- Applying the endpoint representations, we obtain the left- and right- endpoint equations as $2x = 4$ and $3x = 5$.
  - From the left equation, we have $x = 2$ and
  $$[2,3] * 2 = [4,6] \supset [4,5].$$
  The estimation ratio is 50%.
  - From the right equation, we have $x = 5/3$ and
  $$[2,3] * 5/3 = [10/3, 15/3] \supset [4,5],$$
  for an estimation ratio of 60%.
- Using the central point equation $m([2,3])x = m([4,5])$ (i.e., $2.5m(\boldsymbol{x}) = 4.5$, we have $x = 9/5$ and
$$[2,3][9/5,9/5] = [18/5, 27/5] \supset [4,5],$$
for an estimation ratio of
$$\frac{1}{27/5 - 18/5} = 5/9 \approx 55.6\%.$$

The estimation ratios are 60% (right endpoint equation), 55.6% (midpoint equation), 50% (left endpoint equation), and 20% (interval arithmetic). Actually, 60% is the highest estimation ratio among all intervals $\boldsymbol{x}$ that satisfies $[2,3]\boldsymbol{x} \supset [4,5]$. This is because that is the solution of the optimization problem:

Maximize:
$$\frac{5-4}{3\overline{x} - 2\underline{x}}$$
Subject to:
$$\text{(i) } 2\underline{x} \le 4,$$
$$\text{(ii) } 3\overline{x} \ge 5, \text{ and}$$
$$\text{(iii) } \underline{x} \le \overline{x}.$$

In general, the coefficient matrix can be an $n \times n$ matrix and $n > 1$. If we use $\boldsymbol{y}_i$ to represent the interval dot product of the i-th row of $\boldsymbol{A}$ and $\boldsymbol{x}$ (i. e., $\boldsymbol{y}_i = \sum_{1 \le j \le n} \boldsymbol{a}_{ij}\boldsymbol{x}_j$), then finding the "best" approximated solution for $\boldsymbol{A}x = \boldsymbol{b}$, in terms of maximizing the estimation ratio, is a constrained nonlinear optimization problem.

Again, in practice, we may use the endpoints and/or midpoint-width representations as we have done in the above example to find an approximated solution $\boldsymbol{x}$ instead. Such approximated solutions may not be the best (i. e., they may not have the highest possible estimation ratios), but they can be good enough in practice.

The approximated solutions obtained with endpoint or midpoint matrices, if they exist, are degenerate interval vectors, that is, the lower and upper bounds are the same. By adding an $\epsilon > 0$ and/or subtracting it from entries of a degenerated interval vector, we can obtain a nondegenerate interval vector. This process is called an ϵ-inflation. For a degenerate approximated solution, we may perform an ϵ-inflation to obtain an approximated interval vector solution with a higher estimation ratio. We use the following example to illustrate this idea.

Example 4. Find an approximated solution for the interval system of equations $\boldsymbol{A}\boldsymbol{x} = \boldsymbol{b}$ and then improve its quality, where

$$\boldsymbol{A} = \begin{pmatrix} [0,\, 2] & [6,\, 8] & [-2,\, 0] \\ [5,\, 7] & [2,\, 6] & [1,\, 3] \\ [-8,\, -4] & [-1,\, 1] & [1,\, 2] \end{pmatrix},$$

$$\boldsymbol{b} = \begin{pmatrix} [-9,\, 5] \\ [3,\, 9] \\ [-2,\, 4] \end{pmatrix}.$$

To apply the midpoint equation $A_M x = b_M$ for an approximated solution, we calculate

$$A_M = \begin{pmatrix} 1 & 7 & -1 \\ 6 & 4 & 2 \\ -6 & 0 & 1.5 \end{pmatrix}, \text{ and}$$

$$b_M = \begin{pmatrix} -2 \\ 6 \\ 1 \end{pmatrix}.$$

We then obtain the approximated point solution

$$\boldsymbol{x} = \begin{pmatrix} 0.345455 \\ -0.0424242 \\ 2.04848 \end{pmatrix}, \text{ and}$$

$$\boldsymbol{Ax} = \begin{pmatrix} [-4.43636, 0.436364] \\ [3.52121, 8.47879] \\ [-0.757576, 2.75758] \end{pmatrix} \subset \boldsymbol{b}.$$

The solution vector is an inner approximation, since $\boldsymbol{Ax} \subset \boldsymbol{b}$. The estimation ratio is

$$\frac{v(\boldsymbol{Ax})}{v(\boldsymbol{b})} = 0.484073.$$

We use $\epsilon = 0.043$ and perform the ϵ-inflation on both sides of every entry of $\boldsymbol{x}$. Then we have

$$\boldsymbol{x}^* = \begin{pmatrix} [0.302455, 0.388455] \\ [-0.0854242, 0.000575758] \\ [2.00548, 2.09148] \end{pmatrix}.$$

This results in

$$\boldsymbol{Ax}^* = \begin{pmatrix} [-4.86636, 0.781515] \\ [3.00521, 8.99709] \\ [-1.18758, 3.05858] \end{pmatrix} \subset \boldsymbol{b}.$$

The estimation ratio for $\boldsymbol{x}^*$, which is also an inner approximation, is 0.706734.

Certainly, one can perform an ϵ-inflation to selected entries (and/or selected sides with even different ϵ value) of $\boldsymbol{x}$ to obtain approximated solution $\boldsymbol{x}$ for an even higher estimation ratio.

4.4 Singular Value Decomposition for an Interval Matrix

It goes without saying that computational linear algebra plays an essential role in knowledge processing. The concept and algorithms of the singular value decomposition (SVD) for point-valued matrices are among the most significant results in computational linear algebra. To study the SVD for interval matrices, let us first review the definition for a point-valued matrix.

Definition 6. *Let $A = \{a_{ij}\}_{m \times n}$ be an $m \times n$ real matrix. Then there exist an $m \times m$ orthogonal matrix U, an $m \times n$ diagonal matrix Σ with non-negative diagonal entries, and an $n \times n$ orthogonal matrix V such that $A = U\Sigma V^T$. The diagonal elements σ_i, $1 \leq i \leq \min\{m, n\}$, of Σ, are called the singular values of the matrix A; the decomposition is normally computed in such a way that $\sigma_1 \geq \sigma_2 \geq \cdots \geq \sigma_{\min\{m,n\}} \geq 0$. The triplet $\{U, \Sigma, V\}$ is called a* singular value decomposition *(SVD) of A, and the σ_i are called the* singular values *of A.*

The SVD has been one of the most effective tools in solving least squares problems with data, and it is applied in a wide variety of applications, such as control theory, image processing, pattern recognition, time series analysis, and semantic indexing of documents.

We may be tempted to define the SVD for interval matrices as follows.

Definition 7. (Naive definition of the interval SVD) *Let* $\boldsymbol{A} = \{\boldsymbol{a}_{ij}\}_{m \times n}$ *be an* $m \times n$ *interval matrix and let the SVD of a point valued* $m \times n$ *matrix* $A \in \boldsymbol{A}$ *be* $A = U_A \Sigma_A V_A^T$. *The three sets* $\boldsymbol{U} = \{U_A | A \in \boldsymbol{A}, A = U_A \Sigma_A V_A^T\}$, $\Sigma = \{\Sigma_A | A \in \boldsymbol{A}, A = U_A \Sigma_A V_A^T\}$, *and* $\boldsymbol{V} = \{V_A | A \in \boldsymbol{A}, A = U_A \Sigma_A V_A^T\}$ *form a singular value decomposition of* $\boldsymbol{A}$.

There are various problems with the above definition. For instance, the SVD of a point matrix is not unique. For example, if A is the 2×2 identity matrix and U is any orthogonal matrix, then, setting $V = U$, and $\Sigma = I$, $U \Sigma V'$ is an SVD of A. In this case, defining $\boldsymbol{U}$ and $\boldsymbol{V}$ according to Definition 7, the smallest interval matrices $\tilde{\boldsymbol{U}}$ and $\tilde{\boldsymbol{V}}$ that would contain the sets $\boldsymbol{U}$ and $\boldsymbol{V}$ would have $\tilde{\boldsymbol{u}}_{i,j} = \tilde{\boldsymbol{v}}_{i,j} = [-1, 1]$, $i, j = 1, 2$, not a particularly meaningful result. (U can be the identity matrix I, U can be $-I$, U can have 0's on the diagonal and 1's off the diagonal, or U can have 0's on the diagonal and -1's off the diagonal.)

There has been some work at computing the singular values, such as in Deif [5], and Ahn and Chen [1] explain how to compute the exact maximum singular value of an interval matrix. However, computation of bounds $\tilde{\boldsymbol{U}}$ and $\tilde{\boldsymbol{V}}$ for the singular vectors can be more problematical, even in well-behaved cases, such as when each $A \in \boldsymbol{A}$ is symmetric positive definite with distinct eigenvalues. Nonetheless, in knowledge processing, it may not be necessary to compute bounds on the entire sets $\boldsymbol{U}, \boldsymbol{V}$, and $\boldsymbol{\Sigma}$. The endpoints, midpoint, and the width matrices of an interval matrix are point-valued and have special meanings. We can compute an SVD for the A_L, A_R, A_M, and A_W matrices to discover related information. This should reveal useful information about an interval matrix. We use the following example to illustrate this idea.

Example 5. Analyze the behavior of the center and the variation for the interval matrix

$$\boldsymbol{A} = \begin{pmatrix} [0, 2] & [6, 8] & [-2, 0] & [-4, -2] \\ [5, 7] & [2, 6] & [1, 3] & [3, 3] \\ [-8, -4] & [-1, 1] & [0, 0] & [20, 22] \end{pmatrix}.$$

The midpoint-width representation of $\boldsymbol{A}$ is

$$A_M = \begin{pmatrix} 1 & 7 & -1 & -3 \\ 6 & 4 & 2 & 3 \\ -6 & 0 & 0 & 21 \end{pmatrix},$$

$$A_W = \begin{pmatrix} 2 & 2 & 2 & 2 \\ 2 & 4 & 2 & 0 \\ 4 & 2 & 0 & 2 \end{pmatrix}.$$

Computing an SVD for the midpoint (center) matrix with MATLAB, we obtain $A_M = U_M * \Sigma_M * V_M^T$, where

$$U_M \approx \begin{pmatrix} -0.1556 & -0.6053 & 0.7807 \\ 0.0543 & -0.7943 & -0.6051 \\ 0.9863 & -0.0517 & 0.1565 \end{pmatrix},$$

$$\Sigma_M \approx \begin{pmatrix} 22.1217 & 0 & 0 & 0 \\ 0 & 9.1807 & 0 & 0 \\ 0 & 0 & 5.3238 & 0 \end{pmatrix},$$

$$V_M \approx \begin{pmatrix} -0.2598 & -0.5512 & -0.7116 & -0.3496 \\ -0.0394 & -0.8076 & 0.5718 & 0.1387 \\ 0.0119 & -0.1071 & -0.3739 & 0.9212 \\ 0.9648 & -0.1801 & -0.1637 & -0.0999 \end{pmatrix}.$$

From Σ_M, we can see that the most significant singular value related to the midpoint matrix is greater than the sum of the other two.

Similarly, we compute an SVD for the width (variation) matrix and obtain $A_W = U_W * \Sigma_W * V_W^T$, where

$$U_W \approx \begin{pmatrix} -0.5155 & 0.0000 & 0.8569 \\ -0.6059 & -0.7071 & -0.3645 \\ -0.6059 & 0.7071 & -0.3645 \end{pmatrix},$$

$$\Sigma_W \approx \begin{pmatrix} 7.3221 & 0 & 0 & 0 \\ 0 & 2.8284 & 0 & 0 \\ 0 & 0 & 1.5452 & 0 \end{pmatrix},$$

$$V_W \approx \begin{pmatrix} -0.6373 & 0.5000 & -0.3063 & -0.5000 \\ -0.6373 & -0.5000 & -0.3063 & 0.5000 \\ -0.3063 & -0.5000 & 0.6373 & -0.5000 \\ -0.3063 & 0.5000 & 0.6373 & 0.5000 \end{pmatrix}.$$

An important application of the SVD is principal component analysis (PCA), which approximates a general $m \times n$ matrix A by a sum of rank 1 matrices as $A = \sum_{i \le \min\{m,n\}} E_i$, where $E_i = \sigma_i U_i V_i^T$. Using the first principal component to approximate the midpoint matrix A_M, we obtain

$$A_M \approx E_{M_1} \approx \begin{pmatrix} 0.8943 & 0.1357 & -0.0411 & -3.3208 \\ -0.3122 & -0.0474 & 0.0144 & 1.1593 \\ -5.6690 & -0.8599 & 0.2606 & 21.0508 \end{pmatrix}.$$

Similarly, we approximate the width matrix A_W by its first principal component, obtaining

$$A_W \approx E_{W_1} \approx \begin{pmatrix} 2.4056 & 2.4056 & 1.1562 & 1.1562 \\ 2.8275 & 2.8275 & 1.3590 & 1.3590 \\ 2.8275 & 2.8275 & 1.3590 & 1.3590 \end{pmatrix}.$$

Then we can approximate $\boldsymbol{A}$ with $[E_{M_1} - E_{W_1}/2, E_{M_1} + E_{W_1}/2]$ as

$$\boldsymbol{A} \approx \begin{pmatrix} [-0.3085, 2.0971] & [-1.0671, 1.3384] & [-0.6192, 0.5370] & [-3.8988, -2.7427] \\ [-1.7260, 1.1015] & [-1.4611, 1.3664] & [-0.6651, 0.6938] & [0.4799, 1.8388] \\ [-7.0828, -4.2553] & [-2.2737, 0.5538] & [-0.4189, 0.9401] & [20.3713, 21.7303] \end{pmatrix}.$$

This approximation looks significantly different from $\boldsymbol{A}$, but it maintains the most important characteristics of the midpoint and width of $\boldsymbol{A}$. In terms of knowledge processing, the approximation filters out relatively insignificant features of the original data. (We illustrate this with examples in Section 4.5.)

The midpoint and width matrices represent the average and the variability of an interval matrix and related data. Their PCA can be used to form an interval matrix $\tilde{\boldsymbol{A}}$ that approximates the original matrix $\boldsymbol{A}$ but that represents a proper subset of $\boldsymbol{A}$. Therefore, studying the SVD of an interval matrix with its midpoint-width representation is a preferred approach in knowledge representation.

Of course, one may also apply PCA to the endpoint matrices. However, the approximations of endpoint matrices may not necessarily reconstruct an interval matrix. For example, the rank 1 approximation of the left and right endpoint approximations of the above interval matrix $\boldsymbol{A}$ are

$$A_L \approx E_{L_1} \approx \begin{pmatrix} 1.5069 & 0.4112 & -0.0860 & -4.0474 \\ -0.3098 & -0.0845 & 0.0177 & 0.8321 \\ -7.4762 & -2.0401 & 0.4265 & 20.0803 \end{pmatrix}, \text{ and}$$

$$A_R \approx E_{R_1} \approx \begin{pmatrix} 0.3043 & -0.0756 & -0.0251 & -1.9513 \\ -0.3313 & 0.0823 & 0.0273 & 2.1241 \\ -3.4448 & 0.8557 & 0.2836 & 22.0885 \end{pmatrix}.$$

Since there are elements in E_{L_1} that are greater than corresponding entries in E_{R_1}, $[E_{L_1}, E_{R_1}]$ does not represent a valid interval matrix. This gives us another reason for the preference for using the midpoint-width representation in studying the SVD of an interval matrix.

4.5 An Application

In this section, we study an application of the interval matrix SVD to the annual behavior of the S & P 500 index. Chen, Roll, and Ross in [4] state that changes in the overall stock market value (SP_t) are linearly determined by the following five macroeconomic factors: the growth rate variations of seasonally adjusted Industrial Production Index (IP_t), changes in expected inflation (DI_t) and unexpected inflation (UI_t), default risk premiums (DF_t), and unexpected changes in interest rates (TM_t). This relationship can be expressed as

$$SP_t = a_t + I_t(IP_t) + U_t(UI_t) + D_t(DI_t) + F_t(DF_t) + T_t(TM_t).$$

The dataset available for this study consists of monthly point data related to these attributes from January 1930 to December, 2004. The complete dataset is available from the website for this book. Here are the first and the last few sample lines of data:

Yr-mth	UI	DI	SP	IP	DF	TM
30-Jan	-0.00897673	0	0.014382062	-0.003860512	0.0116	-0.0094
30-Feb	-0.00671673	-0.0023	0.060760088	-0.015592832	-0.0057	0.0115
30-Mar	-0.00834673	0.0016	0.037017628	-0.00788855	0.0055	0.0053
30-Apr	0.00295327	0.0005	0.061557893	-0.015966279	0.01	-0.0051
30-May	-0.00744673	-0.0014	-0.061557893	-0.028707502	-0.0082	0.0118
30-Jun	-0.00797673	0.0005	-0.106567965	-0.046763234	0.0059	0.0025
30-Jul	-0.01414673	0.0004	-0.021607229	-0.02193391	0.0022	0.0007
30-Aug	-0.00748673	-0.0007	-0.012903405	-0.017900239	0.0123	-0.0007
30-Sep	0.00569327	-0.0011	-0.000481116	-0.027620585	0.0034	0.0065
30-Oct	-0.00768673	0.0013	-0.148073578	-0.023482607	0.0019	0.0013
30-Nov	-0.00642673	-0.0013	-0.075310618	-0.024047327	-0.0054	0.0033
30-Dec	-0.01924673	0.0004	-0.069121812	-0.004879645	-0.002	-0.0083
...	...	...	...	...	...	...
04-Sep	0.00156327	0.0001	0.026033651	0.007217235	0.0005	0.0085
04-Oct	0.00470327	0	0.000368476	0.002001341	0.001	0.0143
04-Nov	-0.00002273	0	0.044493038	0.006654848	0.0034	-0.0245
04-Dec	-0.00461673	0.0004	0.025567309	0.001918659	0.0007	0.0235

In studying the annual behavior of the stock market, we can reduce the above point matrix to a smaller interval matrix by using the annual minimum and maximum for each attribute. For example, the above point-valued data from January 1930 to December 1930 result in the intervals

$$\begin{aligned} UI_{1930} &= [-0.01924673,\ 0.00569327], \\ DI_{1930} &= [-0.0023,\ 0.0016], \\ SP_{1930} &= [-0.148073578, 0.061557893], \\ IP_{1930} &= [-0.046763234, -0.003860512], \\ DF_{1930} &= [-0.0082, 0.0123], \\ TM_{1930} &= [-0.0094,\ 0.0118]. \end{aligned}$$

Hence, the monthly data from January 1930 to December 2004 form a 75×6 interval matrix $\boldsymbol{A}$, from which we compute the midpoint and width matrices A_M and A_W for the midpoint-width representation of $\boldsymbol{A}$. In studying the behavior of the midpoint of the S & P 500 index during the 75 years between 1930 and 2004, we perform a discrete Fourier analysis on the SP column of A_M. It reveals a strong period of 3.083 years, as is shown in Figure 4.1.

Using MATLAB to compute an SVD for the midpoint matrix A_M, we obtain its singular value matrix:

$$\Sigma_M \approx \begin{pmatrix} 0.2202 & 0 & 0 & 0 & 0 & 0 \\ 0 & 0.1030 & 0 & 0 & 0 & 0 \\ 0 & 0 & 0.0904 & 0 & 0 & 0 \\ 0 & 0 & 0 & 0.0417 & 0 & 0 \\ 0 & 0 & 0 & 0 & 0.0340 & 0 \\ 0 & 0 & 0 & 0 & 0 & 0.0019 \end{pmatrix}.$$

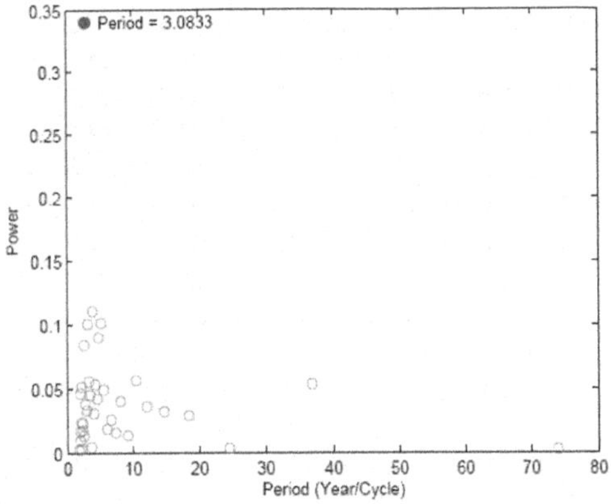

Fig. 4.1. Discrete Fourier analysis for the midpoint matrix for the S & P 500 data from 1930 to 2004.

Since the first singular value is significantly larger than the others, we use the first principal component to approximate the midpoint matrix. Then we perform the discrete Fourier analysis again, but on the SP column of the rank 1 approximation of A_M. It also reveals a strong period of 3.083 years, as is seen in Figure 4.2. Actually, one would have a difficult time finding very significant differences between Figures 4.1 and 4.2. This implies that the first PCA approximates A_M fairly well. However, the calculated 1-norm between the two vectors is about 0.1291.

Similarly, let us investigate the width matrix A_W with PCA. By performing the SVD on the width matrix A_W, we obtain the singular value matrix

$$\Sigma_W \approx \begin{pmatrix} 1.5416 & 0 & 0 & 0 & 0 & 0 \\ 0 & 0.4755 & 0 & 0 & 0 & 0 \\ 0 & 0 & 0.2691 & 0 & 0 & 0 \\ 0 & 0 & 0 & 0.1382 & 0 & 0 \\ 0 & 0 & 0 & 0 & 0.0591 & 0 \\ 0 & 0 & 0 & 0 & 0 & 0.0090 \end{pmatrix}.$$

We use the first principal component to approximate the width matrix A_W, since its first singular value is significantly larger than the others. Figures 4.3 and 4.4 show the results of discrete Fourier analysis on the width of the S & P 500 and its first principal component approximation. Both of them indicate that the annual variation range (width) of the S & P 500 has a significant period of 37 years. However, the 1-norm of the difference between the width vector and its first principal component estimation is 1.1090.

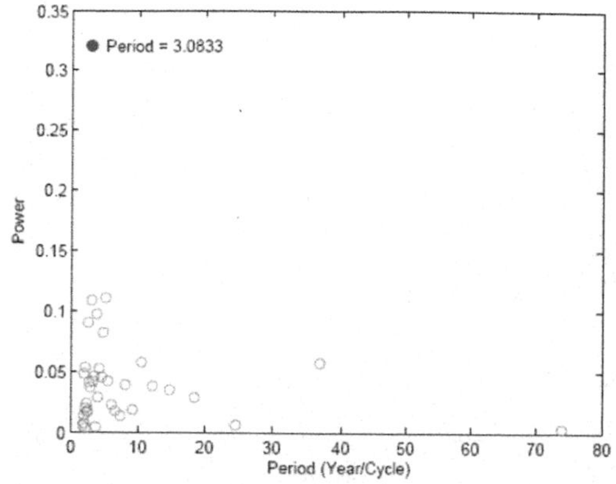

Fig. 4.2. Discrete Fourier analysis for the first PCA of the S & P 500 midpoint matrix

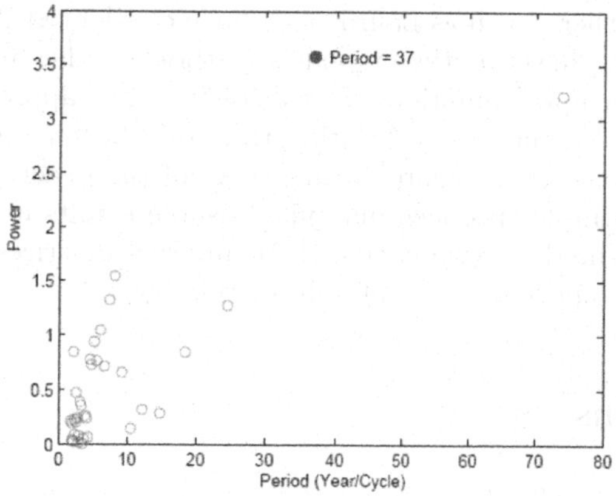

Fig. 4.3. Discrete Fourier analysis for the width matrix for the S & P 500 data from 1930 to 2004.

In this application, we see that a lower-rank PCA approximation can preserve significant information of the midpoint and width matrices of an interval matrix well. By properly combining the information for both midpoint and width matrices, one may obtain related knowledge. Applying PCA to the

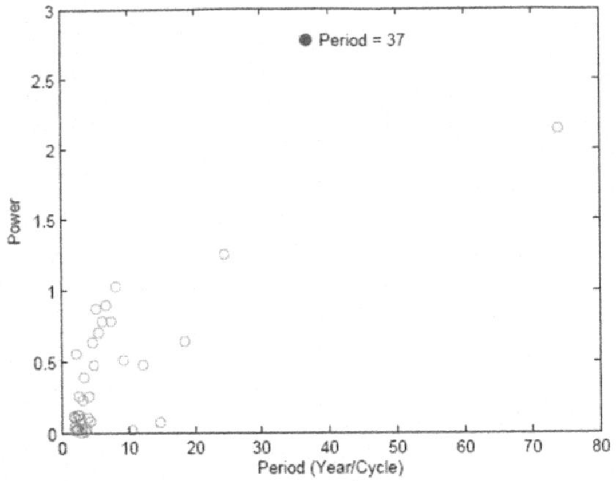

Fig. 4.4. Discrete Fourier analysis for the first PCA of the S & P midpoint matrix

midpoint-width representation of an interval matrix has a great potential in knowledge processing on interval-valued, high-dimensional massive datasets.

We note that others, such as Lauro et al. have considered PCA on interval data in [6] and [8]. However, the emphasis in those studies has been on obtaining sharp outer approximations. In contrast, in our approach, we obtain inner approximations, then use ϵ-inflation to obtain better inner approximations. In applications where rigorous enclosures of the exact solution set are not required, our simple approach may yield usable results in more cases.

The object-oriented environment [11] for interval matrices used to study this application is reported in Chapter 10 of this book.

4.6 Conclusions

In this chapter, we have investigated the usefulness of interval matrices in knowledge processing. The endpoint and midpoint-width representations of an interval matrix reflect the bounds, center, and variance ranges of datasets. Hence, they are important characteristics in knowledge processing.

We have focused our study on interval linear algebra related topics in this chapter. For an interval linear system of equations, in terms of knowledge reliability, we are interested more in finding inner approximated solutions rather than an enclosure containing all point solutions. We have developed the concept of estimation ratio as a useful quality indicator.

The singular value decomposition and principal component analysis on the midpoint and width matrices of an interval matrix can be very powerful tools for information discovery and dimension reduction. This is evidenced by the application to the analysis of the S & P 500 data from 1930 to 2004. The computational results reveal knowledge unknown before.

Acknowledgment: This work is partially supported by the U.S. National Science Foundation under grants CISE/CCF-0202042 and CISE/CCF-0727798. Dr. Ling T. He, Professor of Finance at the University of Central Arkansas, provided the dataset used in Section 4.5.

References

1. Ahn, H.S., Chen, Y.Q.: Exact maximum singular value calculation of an interval matrix. IEEE Transactions on Automatic Control 52, 510–514 (2007)
2. Alefeld, G.: Über Die Durchführbarkeit Des Gaussschen Algorithmus bei Gleichungen mit Intervallen als Koeffizienten. Computing 1 (Supplementum), 15-19 (1977)
3. Chabert, G., Goldsztejn, A.: Extension of the Hansen-Bliek method to right-quantified linear systems. Reliable Computing 13(4), 325–349 (2007)
4. Chen, N.F., Roll, R., Ross, S.A.: Economic forces and the stock market. Journal of Business 59(3), 383–403 (1986). `http://ideas.repec.org/a/ucp/jnlbus/v59y1986i3p383-403.html`
5. Deif, A.S.: Singular values of an interval matrix. Linear Algebra and its Applications 151, 125–133 (1991)
6. Gioia, F., Lauro, C.N.: Principal component analysis on interval data. Computational Statistics 21(2), 343–363 (2006)
7. Kreinovich, V., Lakeyev, A., Rohn, J., Kahl, P.: Computational Complexity and Feasibility of Data Processing and Interval Computations. Applied Optimization Vol. 10. Kluwer Academic Publishers Group, Norwell, MA (1998)
8. Lauro, C., Palumbo, F.: New approaches to principal component analysis of interval data (1998). `citeseer.ist.psu.edu/lauro98new.html`
9. Nash, J.: Equilibrium points in n-person games. Proceedings of the National Academy of Sciences of the United States of America 36, 48–49 (1950)
10. Neumaier, A.: Interval Methods for Systems of Equations, Encyclopedia of Mathematics and its Applications, Vol. 37. Cambridge University Press, Cambridge, UK (1990)
11. Nooner, M., Hu, C.: A computational environment for interval matrices. In: R.L. Muhanna, R.L. Mullen (eds.)Proceedings of a workshop on Reliable Engineering Computing, pp. 65–74. Georgia Tech. University, Savanna, GA (2006) `http://www.gtsav.gatech.edu/workshop/rec06/proceedings.html`

5

Interval Function Approximation and Applications

Chenyi Hu[1], Ling T. He[2], and Shanying Xu[3]

[1] Department of Computer Science, University of Central Arkansas, 201 Donaghey Avenue, Conway, AR 72035-0001, USA. chu@uca.edu
[2] Department of Economics and Finance, University of Central Arkansas, 201 Donaghey Avenue, Conway, AR 72035-0001, USA. linghe@uca.edu
[3] Academy of Mathematics and Systems Science, Chinese Academy of Sciences, Beijing, China. xsy@iss.ac.cn

In this chapter, we use interval functions to describe uncertainties that appear as variabilities of observed data under similar conditions. We then propose practical algorithms to approximate an interval function. By applying the proposed algorithm to actual stock market and crude oil prices data, we obtain quality forecasts as two case studies.

5.1 Introduction

5.1.1 Interval Function

Functions have been among the most studied topics in mathematics and applications. If the analytical form of a function is provided, one can analyze properties and behavior of the function very well. However, in real-world applications, the analytical form of a function is often unknown. Moreover, in using a function $y = f(x)$ to model real-world phenomena, multiple observations of y for a fixed x can often be different. This reflects uncertainties that are traditionally modeled with probability theory as random noise. We notice that the variations of a function value are usually within an interval rather than completely random. Imprecise measurement and control can also cause the value of x to be within an interval rather than an exact point. Therefore, an observation can be recorded as an interval-valued pair $(\boldsymbol{x}, \boldsymbol{y})$.

A real-valued function f can be a *volatile* in a domain D. This means that the sign of its derivative alternates frequently within any small subset of D. Real-world examples of volatile functions include stock prices during a volatile trading day or recorded seismic waves. Figure 5.1 illustrates a volatile function[4]. For such functions, it would be more appropriate to record an obser-

[4] This figure is provided by Mr. Michael Nooner.

C. Hu et al. (eds.), *Knowledge Processing with Interval and Soft Computing*,
DOI: 10.1007/978-1-84800-326-2_5, © Springer-Verlag London Limited 2008

vation with an interval-valued pair $(\boldsymbol{x}, \boldsymbol{y})$ rather than a point. This is mainly because of the following two reasons. First, it would be hard to precisely determine the exact point due to measurement difficulty. Second, even if one can measure discrete point pairs (x, y) precisely, such pairs do not properly reflect the volatile behavior of the function. Hence, such point pairs in classical function approximation can be misleading.

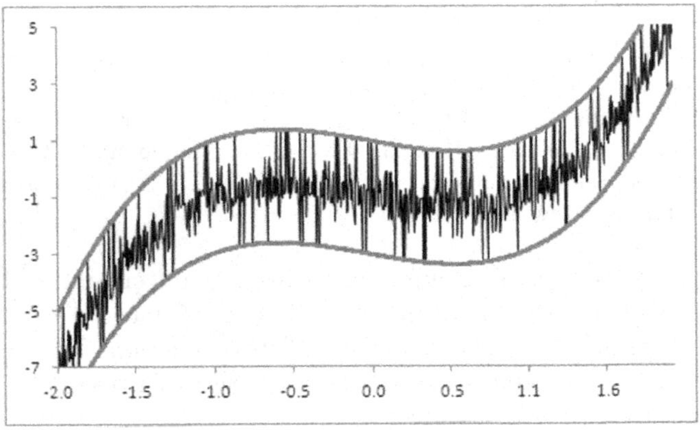

Fig. 5.1. A volatile function.

In this chapter, we view uncertainty as function volatility modeled with interval-valued functions. We extend classical function definition to the concept of *interval function* as follows.

Definition 1. *Let f be a mapping from $\mathbb{R}^n \rightarrow \mathbb{R}$ and let $\vec{\boldsymbol{x}}$ be an interval vector in $\mathbb{R}^n$ (i.e., each component of $\boldsymbol{x}$ is an interval in $\mathbb{R}$). If for any interval vector $\vec{\boldsymbol{x}}$ there is an interval $\boldsymbol{y}$ such that $f(\vec{\boldsymbol{x}}) = \boldsymbol{y}$, then f is an interval function.*

When both $\boldsymbol{x}$ and $\boldsymbol{y}$ are degenerate, an interval function is the same as a classical function. On the other hand, a classical function can also be viewed as an interval function. This is because the range of a function over an $\boldsymbol{x}$ in the domain is usually an interval.

5.1.2 Interval Function Approximation

Using observed discrete data pairs (x, y) to computationally approximate an unknown function has been intensively studied. Our objective here is to establish a general algorithm to approximate an unknown interval function, given a collection of interval-valued pairs $(\boldsymbol{x}, \boldsymbol{y})$. Algorithms for interpolating interval functions have been discussed in [9]. Interval least squares (ILS) function approximation and applications are reported in [8], [10], [7] and [15]. In this chapter, we focus on least squares approximation, since it is probably the most broadly used computational method in function approximation.

5.2 Least Squares Approximation

Let us briefly review the principle and computational methods for least squares approximation.

Definition 2. *Let F be a function space, and let $\Phi = \{\varphi_0, \varphi_1, \dots, \varphi_n, \dots\}$ be a set of functions in F. We say that Φ is a basis for F if for any function $f \in F$ and any given $\epsilon > 0$, there is a linear combination of φ,*

$$f = \sum_j \alpha_j \varphi_j,$$

such that

$$|f(x) - \sum_j \alpha_j \varphi_j| < \epsilon$$

for all x in the domain.

For example, the set $\{1, x, x^2, \dots\}$ is a basis for both polynomial function space and for continuous function space. Commonly used bases for a continuous function space include Chebyshev polynomials, Legendre polynomials, sine/cosine functions, and others.

For a continuous function f (or even if f has a countable number of discontinuities), we may approximate f as

$$f(x) \approx \sum_{0 \le j \le m} \alpha_j \varphi_j(x),$$

where $\{\varphi_j\}_{j=1}^m$ is a preselected set of m basis functions. To determine the coefficient vector $\alpha = (\alpha_0, \alpha_1, \dots, \alpha_m)^T$, the least squares principle requires that the integral of the squares of the differences between $f(x)$ and $\sum_{0 \le j \le m} \alpha_j \varphi_j(x)$ be minimized. In other words, applying the least squares principle to approximating a function f, one selects the vector $\alpha = (\alpha_0, \alpha_1, \cdots, \alpha_m)^T$ that minimizes

$$\int \Big(f(x) - \sum_{0 \le j \le m} \alpha_j \varphi_j(x)\Big)^2 dx.$$

In real-world applications, one usually only knows a collection of N pairs of (x_i, y_i), rather than the function $y = f(x)$. Therefore, one finds an approximation to f by minimizing the following total sum instead:

$$\sum_{i=1}^{N}\left(y_i - \sum_{0\le j\le m} \alpha_j\varphi_j(x_i)\right)^2$$

The classical algorithm that computationally determines the coefficient vector α is as follows.

Algorithm 1 (Classical least squares)

1. *Evaluate the basis functions* $\varphi_j(x)$ *at* x_i *for all* $1 \le i \le N$ *and* $1 \le j \le m$.
2. *Form the matrix*

$$A = \begin{pmatrix} N & \sum_i \varphi_1 & \sum_i \varphi_2 & \cdots \sum_i \varphi_m \\ \sum_i \varphi_1 & \sum_i \varphi_1^2 & \sum_i \varphi_1\varphi_2 & \cdots \sum_i \varphi_1\varphi_m \\ \sum_i \varphi_2 & \sum_i \varphi_2\varphi_1 & \sum_i \varphi_2^2 & \cdots \sum_i \varphi_2\varphi_m \\ \vdots & \vdots & \vdots & \ddots \vdots \\ \sum_i \varphi_m & \sum_i \varphi_m\varphi_1 & \sum_i \varphi_m\varphi_2 & \cdots \sum_i \varphi_m^2 \end{pmatrix} \tag{5.1}$$

and the vector

$$b = \left(\sum_i y_i \ \sum_i y_i\varphi_1(x_i) \ \sum_i y_i\varphi_2(x_i) \ \cdots \sum_i y_i\varphi_m(x_i)\right)^T . \tag{5.2}$$

3. *Solve the linear system of equations* $A\alpha = b$ *for* α.

The above linear system of equations $A\alpha = b$ is called the set of *normal equations*. Instead of the normal equations, a more recent approach applies a design matrix with a sequence of Householder transformations[5] to estimate the vector α. For details concerning Householder transformations, readers may refer to [11] or most books that cover computational linear algebra. Using Householder transformations, least squares function approximation can be described as follows.

Algorithm 2 (Least squares with Householder transformations)

1. *Evaluate the basis functions* $\varphi_j(x)$ *at* x_i *for all* $0 \le i \le n$ *and* $1 \le j \le m$.

[5] That is, modern point algorithms use a QR factorization or a singular value decomposition to avoid ill-conditioning of the matrix (5.1), whereas the normal equations are mostly of theoretical interest.

2. *Form the design matrix*

$$X = \begin{pmatrix} \varphi_1(x_1) & \varphi_2(x_1) & \cdots & \varphi_m(x_1) \\ \varphi_1(x_2) & \varphi_2(x_2) & \cdots & \varphi_m(x_2) \\ \cdots & \cdots & \cdots & \cdots \\ \varphi_1(x_i) & \varphi_2(x_i) & \cdots & \varphi_m(x_i) \\ \cdots & \cdots & \cdots & \cdots \\ \varphi_1(x_N) & \varphi_2(x_N) & \cdots & \varphi_m(x_N) \end{pmatrix}. \tag{5.3}$$

3. *Perform a sequence of Householder transformations to X to produce an upper triangular matrix R;*
4. *Apply the same sequence of Householder transformations, in the same order, to the vector $(y_1, y_2, \ldots, y_N)^T$ and obtain a vector z.*
5. *Solve $R\alpha = z$ for α.*

Computing a point least squares approximation using either Algorithm 1 or Algorithm 2 is called *ordinary least squares* (OLS) approximation.

5.3 Interval Function Approximation

Most, if not all, previous studies on least squares approximation assume point-valued data. There are several computational issues that need to be considered to apply the above algorithms to interval-valued pairs $(\boldsymbol{x}, \boldsymbol{y})$ to approximate an interval function.

5.3.1 Computational Challenges

With interval arithmetic [12], it is straightforward to perform both steps 1 and 2 in Algorithm 1. However, it presents a challenge in step 3. This is because the normal equations are now interval systems of linear equations $\boldsymbol{A}\alpha = \boldsymbol{b}$. The solution set of an interval linear system of equations is mostly irregularly shaped and nonconvex [13]. A naive application of interval arithmetic to bound the solution vector α may cause serious overestimation. Using the design matrix approach would not solve the problem, since finding a Householder transformation for an interval matrix remains a challenge.

5.3.2 An Inner Approximation Approach

As discussed in the previous chapter, an interval $\boldsymbol{x} = [\underline{x}, \overline{x}]$ can be represented by its midpoint $\text{mid}(\boldsymbol{x}) = (\underline{x} + \overline{x})/2$ and its width $w(\boldsymbol{x}) = \overline{x} - \underline{x}$. We can take a two-step approach of considering the midpoint and width separately to determine α. In the step 3 of Algorithm 1, let us first find the midpoint vector of α, which is a point vector. This suggests matching the center of the two interval vectors $\boldsymbol{A}\alpha$ and $\boldsymbol{b}$ in the interval linear system of equations $\boldsymbol{A}\alpha = \boldsymbol{b}$.

Let A_{mid} be the midpoint matrix of $\boldsymbol{A}$ and let b_{mid} be the midpoint vector of $\boldsymbol{b}$. We solve the linear system of equations $A_{mid}\alpha = b_{mid}$ for α.

Similarly, the midpoint matrix of interval matrix (5.3) is a point matrix X_{mid}. Hence, we can perform a sequence of Householder transformation as required by step 3 of Algorithm 2 to obtain a point upper triangular matrix R. The sequence of Householder transformations can be applied directly to the interval vector $(\boldsymbol{y}_1, \boldsymbol{y}_2, \cdots, \boldsymbol{y}_N)^T$ as in step 4 of Algorithm 2 to obtain an interval vector $\boldsymbol{z}$. In step 5 of Algorithm 2, we solve $R\alpha = z_{mid}$ to obtain an approximation for the midpoint of the interval vector α.

We emphasize that the result of evaluating

$$\boldsymbol{y} = f(\boldsymbol{x}) \approx \alpha_0 + \sum_{1 \le j \le m} \alpha_j \varphi_j(\boldsymbol{x})$$

is an interval even if we use the midpoint of the interval vector α in the calculation. This is because the independent variable $\boldsymbol{x}$ is interval-valued. However, by collapsing an interval vector α to its midpoint, we may expect that the initial approximation is an *inner interval approximation.*

5.3.3 Width Adjustment

Now let us consider ways to determine the width of $\boldsymbol{y}$. There can be a number of different computational heuristics. One of them is to perform ϵ-inflation, as reported in Section 4.3 of Chapter 4, on the point-valued α. By maximizing the estimation ratio of $\boldsymbol{A}\alpha = \boldsymbol{b}$ in Algorithm 1 (or $R\alpha = \boldsymbol{z}$ in Algorithm 2), we obtain an interval-valued α. Another approach is to estimate the width of $\boldsymbol{y}$ and adjust the inner approximation directly by multiplying a scale factor. For example, we can apply least squares approximation to estimate the width of $\boldsymbol{y}$. There are open questions that need to be further studied on optimal width adjustment.

5.3.4 Interval Least Squares Approximation

Summarizing the above discussion, the main steps of an interval function least squares approximation algorithm include the following:

1. Input available interval data pairs $(\boldsymbol{x}_i, \boldsymbol{y}_i)$.
2. Use interval arithmetic to evaluate the normal equations or the design matrix.
3. Find an initial approximation.
4. Modify the initial approximation with a width adjustment.

We apply these steps in the case studies in this chapter.

5.3.5 Other Approaches to Obtain an Interval Approximation

One may obtain an interval approximation without using interval arithmetic at all. The lower and upper bounds of interval data pairs $(\boldsymbol{x}_i, \boldsymbol{y}_i)$ form two collections of point data $(\underline{x}_i, \underline{y}_i)$ and $(\overline{x}_i, \overline{y}_i)$. By applying point least squares approximation separately to these two collections, one can obtain two point estimations. These two estimations form an interval. We call this approach the *min-max interval approximation.* This has been reported and applied in [6] and [7].

Another way to obtain an interval approximation is to apply classical statistical/probabilistic approaches. By adding to and subtracting from a point approximation a certain percentage of standard deviations, one can obtain forecasting intervals. In the literature, this is called a *confidence interval.* However, our case studies imply that, at least in some cases, interval function least squares approximation can produce better quality computational results than that obtained with the min-max intervals and confidence intervals.

5.4 Assessing Interval Function Approximation

There are different ways to produce an interval function approximation. An immediate question is how to assess the quality. We define two measurements, - *absolute error* and *accuracy ratio* -, to assess an interval approximation defined as follows.

Definition 3. *Let $\boldsymbol{y}_{est}$ be an approximation for the interval $\boldsymbol{y}$. The absolute error of the approximation is the absolute sum of the lower and upper bound errors, that is,*

$$|\underline{y}_{est} - \underline{y}| + |\overline{y}_{est} - \overline{y}|.$$

Example 1. If one uses $[-1.02, 1.95]$ to approximate the interval $[-1.0, 2.0]$, then the absolute error of the estimation is

$$|(-1.02) - (-1.0)| + |1.95 - 2.0| = 0.02 + 0.05 = 0.07.$$

Since both $\boldsymbol{y}_{\text{est}}$ and $\boldsymbol{y}$ are intervals, an additional meaningful measure of quality can be defined. The larger the overlap between the two intervals, the better the approximation should be. By the same token, the less the nonoverlap between the two intervals, the more accurate the forecast is. In addition, the accuracy of an interval estimation should be between 0% and 100%. By using the notion of interval width, which is the difference between the upper and lower bounds of an interval, we can measure the intersection and the union (or the convex hull) of the two intervals. The function $w(\ )$ returns the width of an interval. We define the concept, named the *accuracy ratio of an interval approximation,* as follows.

Definition 4. *Let $\boldsymbol{y}_{est}$ be an approximation for the interval $\boldsymbol{y}$. The accuracy ratio of the approximation is*

$$\begin{cases} \dfrac{w(\boldsymbol{y} \cap \boldsymbol{y}_{est})}{w(\boldsymbol{y} \cup \boldsymbol{y}_{est})} & \text{if } (\boldsymbol{y} \cap \boldsymbol{y}_{est}) \neq \emptyset \\ 0 & \text{otherwise.} \end{cases}$$

Example 2. Using $[-1.02, 1.95]$ to approximate the interval $[-1.0, 2.0]$, the accuracy ratio is

$$\frac{w([-1.02, 1.95] \cap [-1.0, 2.0])}{w([-1.02, 1.95] \cup [-1.0, 2.0])} = \frac{w([-1.0, 1.95])}{w([-1.02, 2.0])} = \frac{2.95}{3.02} = 97.68\%.$$

As in classical statistics, one can calculate the mean, standard deviation of a collection of interval estimations, as well as their absolute errors and accuracy ratios. Furthermore, one may compare the overall quality of different approximations.

5.5 Application 1: S & P 500 Index Interval Forecasting

The S & P 500 index is broadly used as a indicator for the overall stock market. Driven by macroeconomic and social factors, the stock market usually varies with time. The main challenge in studying the stock market is its volatility and uncertainty. With interval least squares approximation for the S & P 500 annual interval forecast, we have obtained astonishing computational results reported in [5] and [10] in addition to that reported in Section 4.5 of Chapter 4.

5.5.1 The Model

Arbitrage pricing theory (APT) [14] provides a framework that identifies macroeconomic variables that significantly and systematically influence stock prices. By modeling the relationship between the stock market and relevant macroeconomic variables, one may forecast the overall level of the stock market. The model established by Chen, Roll, and Ross [1] is broadly accepted. According to their model, the changes in the overall stock market value (SP_t) are linearly determined by the following five macroeconomic factors: the growth rate variations of seasonally adjusted Industrial Production Index (IP_t), changes in expected (DI_t) and unexpected (UI_t) inflation, default risk premiums (DF_t), and unexpected changes in interest rates (TM_t). This relationship can be expressed as

$$SP_t = a_t + I_t(IP_t) + U_t(UI_t) + D_t(DI_t) + F_t(DF_t) + T_t(TM_t).$$

By using historic data, one may estimate the coefficients of the above equation to forecast changes of the overall stock market.

5.5.2 Time Series and Slicing-Window

Chronologically ordered historical data form a time series. There is a general consensus in the financial literature that relationships between financial market and macroeconomic variables are time-varying. This means that the relationship is valid only for a limited time period. Therefore, in applying function approximation on a time series, one should use only data inside an appropriate time window to estimate the relationship.

Time series have been extensively studied for prediction and forecasting [3] and [4]. In the literature, it is called an *in-sample forecast* if the coefficients in a time window and the equation above are used to calculate the SP for the last time period in the time window. It is called an *out-of-sample forecast* if the coefficients in a time window and the equation above are used to calculate the SP for the first time period that immediately follows the time-window [2].

By slicing the time window (also called "rolling"), one obtains a sequence of coefficients and forecasted SP values. The overall quality of forecasting can be measured by comparing the forecasts against actual SP values. In practice, the out-of-sample forecast is more useful than in-sample forecast because it can make predictions.

5.5.3 The Data

To date, the primary measurements used in economics and finance are quantified points. For instance, a monthly closing value of an index is used to represent the index for that month, even though the index actually varies during that month. The data used in this case study are monthly data from January 1930 to December 2004. We list a portion of the data here.

Yr_mth	UI	DI	SP	IP	DF	TM
30-Jan	-0.00897673	0	0.014382062	-0.003860512	0.0116	-0.0094
30-Feb	-0.00671673	-0.0023	0.060760088	-0.015592832	-0.0057	0.0115
30-Mar	-0.00834673	0.0016	0.037017628	-0.00788855	0.0055	0.0053
30-Apr	0.00295327	0.0005	0.061557893	-0.015966279	0.01	-0.0051
30-May	-0.00744673	-0.0014	-0.061557893	-0.028707502	-0.0082	0.0118
30-Jun	-0.00797673	0.0005	-0.106567965	-0.046763234	0.0059	0.0025
...	......	...	...	...	...	...
04-Jun	0.00312327	-0.0002	0.026818986	0.005903385	-0.0028	0.0115
04-Jul	-0.00182673	0.0002	-0.024043354	0.00306212	0.0029	0.0147
04-Aug	0.00008127	0.0002	-0.015411102	-0.002424198	0	0.0385
04-Sep	0.00156327	0.0001	0.026033651	0.007217235	0.0005	0.0085
04-Oct	0.00470327	0	0.000368476	0.002001341	0.001	0.0143
04-Nov	-0.00002273	0	0.044493038	0.006654848	0.0034	-0.0245
04-Dec	-0.00461673	0.0004	0.025567309	0.001918659	0.0007	0.0235

In this case study, we use a time window of 10 years to obtain the out-of-sample annual forecasts for 1940-2004.

5.5.4 Interval Rolling Least Squares Forecasts

To perform interval rolling least squares forecasts, we need interval input data. We choose the annual minimum and maximum for each of the attributes of the provided monthly data to form the interval input data. We use the normal equations to obtain an inner approximation and then adjust its width to the average width of the S & P 500 intervals within the time-window. Use the software toolbox reported in Chapter 10; we developed a program that implemented interval function approximation algorithms discussed in this chapter to perform stock market forecasts. Figure 5.2 illustrates the out-of-sample annual interval forecasts.

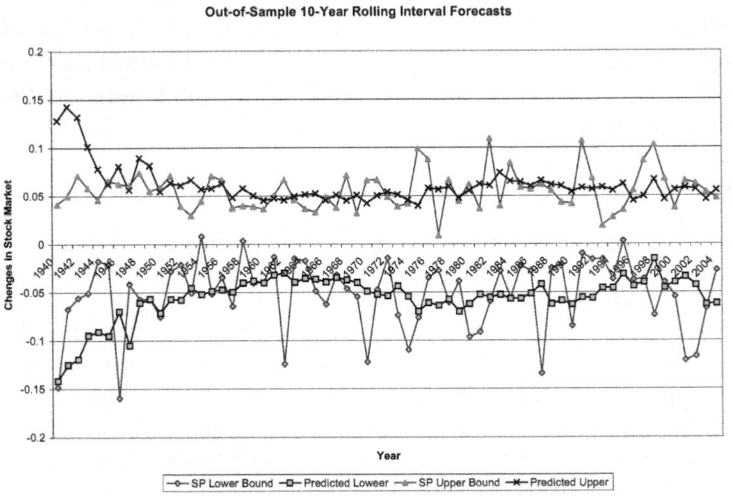

Fig. 5.2. Out-of-sample least squares annual interval forecasts (1940-2004).

5.5.5 Min-Max Interval Forecasts

By applying point rolling OLS to the annual minimum and maximum, we obtain two out-of-sample annual forecasts for the period of 1940-2004. They form annual min-max interval forecasts. Figure 5.3 illustrates the min-max interval forecasts. The time window is ten years, the same as that used in the interval least squares forecast.

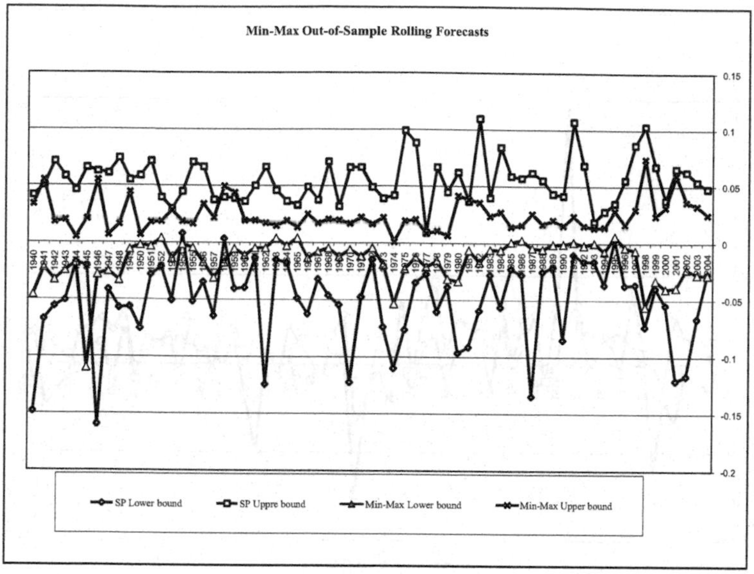

Fig. 5.3. Out-of-sample min-max annual interval forecasts (1940-2004).

5.5.6 Point and Confidence Interval Forecasts

For the purpose of quality comparison, we calculated the annual point forecasts that are commonly used in financial study. We obtained the out-of-sample annual forecasts (in percent) for the period 1940-2004. The out-of-sample annual point forecasts have an average absolute forecasting error of 20.6%, with a standard deviation of 0.19. By adding to and subtracting from the point-forecasts with a proportion of the standard deviation, we may form confidence interval forecasts with 95% statistical confidences. It is worth pointing out that the ranges of interval forecasts are significantly less than that of point forecasts at the ratio of only about 14%.

5.5.7 Quality Comparisons

To assess the quality of the above forecasts, we use the following indicators: (1) the average absolute error, (2) the standard deviation of forecast errors, (3) the average accuracy ratio, and (4) the number of forecasts with 0% accuracy. We summarize the statistics of the quality indicators in Table 5.1.

All measured indicators for forecasting quality in Table 5.1 suggest that interval OLS significantly outperforms point-based forecasts with a much less

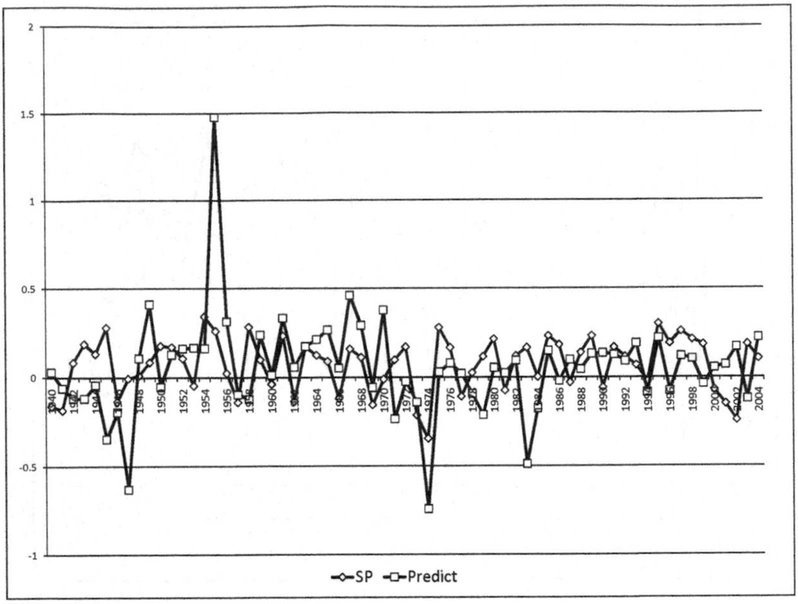

Fig. 5.4. Out-of-sample annual point forecasts (1940-2004).

	Absolute mean error	Standard deviation	Accuracy ratio	Zero accuracy
OLS	0.20572	0.18996	NA	NA
Std dev., 95% confidence	0.72351	0.31197	0.1257	5
Min-Max interval	0.06664	0.04100	0.4617	0
Initial interval Fcast	0.07304	0.03815	0.3855	0
Interval Fcast	0.05166	0.03224	0.6419	0

Table 5.1. Quality Comparison of Annual Forecasts (1940-2004)

mean forecast error. The much smaller standard deviations produced by the interval approaches indicate that the interval forecasting is more stable than other comparing methods. Compared with the point-based confidence interval forecasting, interval methods produce a much higher average accuracy ratio. The interval scheme with width adjustments further improves the overall forecasting quality of initial approximations in terms of the higher average accuracy ratio. All forecasts with interval computing have a positive accuracy ratio, whereas some of the point-based confidence intervals have zero accuracy. In the min-max approach, three predicted lower bounds were greater than the computed upper bounds. We reordered them into proper intervals.

5.6 Application 2: Crude Oil Price Interval Forecasting

As a strategic resource, crude oil and its trade have attracted extensive attention. Crude oil forecasting has been focused on the relationship between commodity inventory levels and spot prices. Previous work has explained a relationship between crude oil price and total inventories (crude plus products) of the Organization for Economic Cooperation and Development (OECD) countries. Since crude oil prices vary frequently within an interval during a given month, we study them with interval forecasting.

5.6.1 The Model

In this study, we use the oil price forecasting model established by Ye, Zyren, and Shore [16]. In that model, the observed level of petroleum inventory is decomposed into two components: the normal level and the relative level. The former is determined by historical seasonal movements and trends and reflects the normal market demand and operational requirements. The later represents the difference between the observed and normal levels and reflects short-run market fluctuations.

Using IN_t and IN^* to represent the actual and normal inventory levels at time t, respectively, the relative inventory level (denoted by RIN) is defined as

$$RIN_t = IN_t - IN_t^*.$$

The normal inventory level is modeled with

$$IN_t^* = a_0 + b_1 T + \sum_{k=2}^{12} b_k D_k \tag{5.4}$$

where D_k for $k = 2, 3, \ldots, 12$ are 11 seasonal dummy variables and T is time, beginning with 1. The coefficients a_0 and b_k for $k = 1, 2, \ldots, 12$ need to be determined.

In [16], Ye et al. claimed that the best specification found for the short-run forecast of the monthly WTI spot price is the Relative Stock (RSTK) model

$$WTI_t = a + \left\{ \sum_{i=0}^{3} b_i RIN_{t-i} \right\} + \left\{ \sum_{j=0}^{5} c_j D_j 911 \right\} + dLAPR99 + eWTI_{t-1} + \epsilon_t, \tag{5.5}$$

where a, b_i, c_j, d, and e are to be determined parameters, $D_j 911$ is a set of single monthly variables to account for market disequilibrium following the September 11, 2001 terrorist attacks in the United States, and $LAPR99$ is a level shifting variable corresponding to the effect that OPEC quota tightening had on the petroleum market beginning in April 1999.

5.6.2 The Data

The data used in this study are monthly averages of West Texas Intermediate crude spot prices and the monthly OECD total inventory levels. They are available from the website of the Energy Information Association (EIA). The monthly average is commonly used in forecasting monthly oil prices. For example, the EIA official website releases monthly data as the average of daily prices in a single month. The WTI spot price is in nominal dollars per barrel, and the inventory level is measured in millions of barrels. To compare with the results reported by Ye et al. in [16] and [17], we study the same period as theirs: from January 1992 to April 2003. The size of the rolling time window is 96 months, the same as theirs. The crude oil spot prices fell and rose dramatically after September 11, 2001. By introducing six dummy variables, the model does not forecast for the 6 months immediately after that event.

We form monthly price intervals by taking the minimum and the maximum of daily prices in the month. The relative inventory level is calculated by using above formula. Given the common sense that the inventories vary during a month, we transform the relative level into interval data by adding and subtracting a certain percentage. Through experiments, we set the percentage to be around 40%, so that the interval least squares approximation produces results comparable with that of 95% confidence interval forecasts.

5.6.3 Computational Results

To examine the quality of the forecasting intervals produced by interval least-squares approximation, we compare them against the 95% confidence interval forecasts based on the point forecasts in [16] and [17]. Table 5.2 lists the average accuracy and average error for both in-sample and out-of-sample forecasts against the actual price intervals.

	Average absolute error	Average accuracy ratio
In-sample		
Interval Forecast	3.147	51.932%
95% conf. interval	3.163	53.239%
Out-of-Sample		
Interval Forecast	3.511	48.557%
95% conf. interval	3.482	50.241

Table 5.2. Least Squares Forecasts vs. 95% Confidence Intervals

In terms of average absolute error and average accuracy ratio, the interval least squares approximation provides only comparable interval forecasts with 95% confidence intervals.

In decision making, the monthly average price is a useful parameter. Therefore, we examined if the monthly average fell within the monthly forecasted interval or not. We found that the interval least squares approximation slightly outperformed 95% confidence intervals. Whereas 82.35% of the average monthly prices fell within the intervals produced by interval least-squares approximation, only 79.41% fell within the 95% confidence intervals.

5.7 Conclusions

In this chapter, we apply interval functions to model uncertainty and volatility. With interval-valued nodes, we may fit an interval function with least squares approximation. The strategy proposed in this chapter is to first find an inner approximation, and then to adjust the width with computational heuristics. Although the interval function to be approximated is unknown, we can still statistically assess the quality of an interval approximation.

We examined two case studies in this chapter. The first deals with annual interval forecasts for the S & P 500 index from 1940 to 2004. Interval function least squares approximation produced significantly better results than that obtained with traditional point approaches in terms of overall less mean error and higher average accuracy ratio. The other application deals with forecasting monthly crude oil price intervals from January 2000 to April 2003. The forecasting intervals obtained with the interval least squares approach are comparable with that of 95% confidence intervals. Although these are initial attempts at using interval methods in financial study, our computational results imply that interval functions and their least square approximation have a good potential in studying volatile phenomena in general.

Acknowledgment: This work is partially supported by the U.S. National Science Foundation under grants CISE/CCF-0202042 and CISE/CCF-0727798.

References

1. Chen, N.F., Roll, R., Ross, S.A.: Economic forces and the stock market. Journal of Business 59(3), 383–403 (1986). `http://ideas.repec.org/a/ucp/jnlbus/v59y1986i3p383-403.html`
2. Fama, E., French, K.: Industry costs of equity. Financial Economics 43, 153–193 (1997)
3. Gardner, E.: A simple method of computing prediction intervals for time series forecasts. Management Science 34, 541–546 (1988)
4. Gooijer, J., Hyndman, R.: 25 years of time series forecasting. Forecasting 22, 443–473 (2006)
5. He, L., Hu, C.: The stock market forecasting: An application of the interval measurement and computation. In: The 2nd International Conference on Fuzzy Sets and Soft Computing in Economics and Finance, pp. 13–22, St. Petersburg (2006)

6. He, L., Hu, C.: Impacts of interval measurement on studies of economic variability: evidence from stock market variability forecasting. Risk Finance 12, 489-507 (2007)
7. He, L., Hu, C.: Impacts of interval computing on stock market variability forecasting. Computational Economics (to appear)
8. Hu, C.: Using interval function approximation to estimate uncertainty. Springer book series: Advances in Soft Computing, vol. 46, Interval / Probabilistic Uncertainty and Non-Classical Logics, 341–352, (2008)
9. Hu, C., Cardenas, A., Hoogendoorn, S., Selpulveda, P.: An interval polynomial interpolation problem and its Lagrange solution. Reliable Computing 4(1), 27–38 (1998)
10. Hu, C., He, L.: An application of interval methods to the stock market forecasting. Reliable Computing 13, 423-434 (2007)
11. Moler, C.B.: Numerical Computing with MATLAB. Society for Industrial and Applied Mathematics, Philadelphia (2004)
12. Moore, R.E.: Methods and Applications of Interval Analysis. Society for Industrial and Applied Mathematics, Philadelphia (1979)
13. Neumaier, A.: Interval Methods for Systems of Equations, Encyclopedia of Mathematics and its Applications, Vol. 37. Cambridge University Press, Cambridge (1990)
14. Ross, S.: The arbitrage theory of capital asset pricing. Economic Theory 13, 341–360 (1976)
15. Xu, S., Chen, X., Ai, H.: Interval forecasting of crude oil price. Springer book series: Advances in Soft Computing, vol. 46, Interval/Probabilistic Uncertainty and Non-Classical Logics, 353-363 (2008)
16. Ye, M., Zyren, J., Shore, J.: A monthly crude oil spot price forecasting model using relative inventories. Forecasting 21, 491–501 (2005)
17. Ye, M., Zyren, J., Shore, J.: Forecasting short-run crude oil price using high and low-inventory variables. Energy Policy 34, 2736–2743 (2006)

6

Interval Rule Matrices for Decision Making

Chenyi Hu

Department of Computer Science, University of Central Arkansas, 201 Donaghey Avenue, Conway, AR 72035-0001, USA. chu@uca.edu

In this chapter, we present a decision-making system using an interval rule matrix. Section 6.2 introduces the rule matrix model. Section 6.3 reports practical algorithms that establish an interval rule matrix. Section 6.4 describes how to use an interval rule matrix to make decisions according to environmental observations.

6.1 Introduction

Every day, numerous decisions are made for various reasons. Although many of these decisions are astute enough and return results as expected, there are others that can be made better to avoid unwanted consequences. How to make a good, a better, or even the best decision is a practical question asked frequently.

A decision problem usually consists of a set of possible states of nature (environment), a set of possible actions (decisions), and a benefit function that measures the results of a decision in a given environment. To make a right decision, one certainly needs to know the current environment and related knowledge. Through matching input data (relevance of each environment feature) with a certain set of rules, one may make a decision accordingly. However, due to uncertainty and incomplete information, the best one can do is to estimate the benefit for a decision. Usually, the actual best decision can only be known afterward. Therefore, historical data are often useful for people to discover rules for decision making.

As the world becomes more complex, the number of features involved in decision making can be unmanageable for human beings. Modern computers collect massive datasets and perform trillions of calculations in seconds. Nowadays, knowledge-based agents are designed, implemented, and embedded into computers as automated decision-making systems. These systems apply decision theories and algorithms to generate rules based on statistical/probabilistic, stochastic, fuzzy systems [5, 11, 12, 15], neuro-fuzzy systems

C. Hu et al. (eds.), *Knowledge Processing with Interval and Soft Computing*, DOI: 10.1007/978-1-84800-326-2_6,

[9], and so on. In this chapter, we specifically study an interval rule matrix model for automated decision making and reasoning about possible courses of actions based on an environmental observation.

6.2 The Rule Matrix Model

6.2.1 A Simple Example

Decisions are mostly made according to observations of the current environment and existing knowledge. Here is a simple example.

Example 1. Bob needs to decide if he should carry his umbrella and coat before leaving home for work. He may carry (a) both his umbrella and coat, (b) his umbrella but not his coat, (c) his coat but not an umbrella, and (d) neither. Unfortunately, he has no way to check the forecasts and must make a quick decision by observing the current weather conditions. Based on his knowledge, Bob uses the following table to make his decision.

Environment/decision	a	b	c	d
Rain or very likely	Yes	Yes	No	No
Current temperature	Below 40°F	Above 40°F	Below 40°F	Above 40°F

The first row of the table lists all possible decisions for Bob and the first column lists environmental parameters that Bob takes into consideration for decision making. By matching the environmental observations with the table, Bob can easily make his decision.

6.2.2 Rule Matrices

In general, an environment e may contain m features, and there can be n possible different decisions, $d_1, d_2, \ldots, d_n$, that could be made based on the presence of the environment features. Let $e = (e_1, e_2, \ldots, e_m)^T$ be an observation of the environment; that is, e denotes the degree to which certain features of an environment are present. Then a knowledge-based agent may select a specific decision by matching the input e with column vectors of the m by n matrix P:

$$P = \begin{pmatrix} p_{11} & p_{12} & \cdots & p_{1n} \\ p_{21} & p_{22} & \cdots & p_{2n} \\ & & \vdots & \\ p_{m1} & p_{m2} & \cdots & p_{mn} \end{pmatrix}.$$

If an observation vector e matches P_j, (i. e., the j-th column of P), then the jth decision d_j should be selected. A benefit function can be associated with the decision as well. We call P a *rule matrix* for the decision-making process, and the model is called a rule matrix model.

In practice, a rule matrix should be interval-valued, as suggested by de Korvin, Hu, and Chen in [4]. This is mainly because of the following two reasons. The first one is that a decision is usually selected according to ranges of attribute values rather than matching a point exactly. In addition, an environment observation often imprecise. A rule matrix generated from such imperfect data should allow an error bounds. By specifying the lower and upper bounds of feature presence in an interval rule matrix, we take uncertainty into consideration appropriately [3, 4, 14].

We denote an interval rule matrix by the boldface uppercase letter $\boldsymbol{P}$ whose entry $\boldsymbol{p}_{ij}$ is an interval for each $1 \leq i \leq m$ and $1 \leq j \leq n$. By the same token, an environment observation is also interval-valued and is denoted by boldface as $\boldsymbol{e}$.

6.2.3 Another Example

For a more sophisticated example, we describe an interval rule matrix to enhance network intrusion detection systems (IDSs). In addition to user authentication, cryptography, and firewalls, IDSs are embedded into network management systems to monitor network states and perform control functions.

Two main types of intrusion detection approach are *misuse detection* [8] and *anomaly detection* [13]. Anomaly detection compares activities with established normal usage patterns (profiles) and determines if the network state deviates over some threshold from normal patterns. This approach is recognized as being capable of catching new attacks. However, normal patterns of usage and system behaviors can vary wildly. Therefore, anomaly detection usually produces a very large number of false alarms, see [2] and [10] and hence is impractical. By representing normal patterns in an interval-valued matrix, we should be able to reduce false alarms caused by a small change in normal behavior or a slight deviation from a pattern derived from the network audit data.

To generate an interval rule matrix for intrusion detection, we may assume the availability of previous network state data and intrusions associated with them. These can be automatically collected by network monitoring software. If we consider m features and n possible interval-valued network states, then we have an $m \times n$ interval rule matrix. We represent the likelihood of an intrusion with 0 (certainly not) and 1 (absolutely yes). Then the likelihood of an intrusion can be represented as a mapping from the network state to the interval $[0, 1]$.

To simplify the example, in [6], Duan, Hu, and Wei considered only three network state attributes: the average packet delay for a Transmission Control Protocol (TCP) flow, the bandwidth utilized for the TCP flow, and the number of TCP flows arriving at one router ingress port (shown as the first, second, and the third rows of the following matrix, respectively). An artificial interval matrix for the network state can be as follows:

$$\begin{pmatrix} [0.8, 1.8] & [1.6, 2.1] & [2.2, 3.1] \\ [3.5, 5.2] & [4.8, 6.5] & [6.9, 7.4] \\ [8, 10] & [9, 12] & [13, 16] \end{pmatrix}.$$

By counting intrusions associated with each of these states, we can obtain an empirical probability of intrusions d_j associated with the j-th column. For a current network state observation $\boldsymbol{e}$, we can estimate the likelihood of a intrusion (or intrusions) as follows:

- *Input:* Current network state $\boldsymbol{e}$.
- *Estimation:* Compare $\boldsymbol{e}$ and column vectors of $\boldsymbol{P}$.
  Case 1: If $\boldsymbol{e} \subseteq \boldsymbol{P}_j$, the likelihood of an intrusion is d_j.
  Case 2: If $\forall j \in \{1, 2, \ldots, n\}$, $\boldsymbol{e} \cap \boldsymbol{P}_j = \emptyset$, then the network state is very abnormal when compared to historical states. An intrusion alarm should be sent to the network administrator.
  Case 3: If $\boldsymbol{e} \cap \boldsymbol{P}_j \neq \emptyset$ for more than one $j \in \{1, 2, \ldots, n\}$, then one may select the greatest d_j or use other heuristics.

By using the empirical probability and cost function, the system can estimate the expected average damage for a possible intrusion. In [6], Duan, Hu, and Wei further suggested that the interval rule matrix model could be integrated with data collection, policy generation, and policy application. According to the current likelihood of network intrusion, network control can then identify risk levels and automatically take actions.

The example constructed earlier shows the potential of interval rule matrix in various kinds of application. In the rest of this chapter, we discuss ways to establish an interval rule matrix and to how apply it to decision making.

6.3 Establishing an Interval-Valued Rule Matrix

The main purpose of this section is to design practical algorithms that construct an interval rule matrix $\boldsymbol{P}$ from a known dataset E. We assume that E contains N environment-decision pairs. Because we have used e_k to indicate the kth feature of an environment, we use a superscript e^k to denote the kth observation of the environment. By the term environment-decision pair $[e^k, d_{k^*}] \in E$, we mean that under a given environment e^k, $1 \leq k \leq N$, the *desired* decision should be d_{k^*} for a particular k^* with $1 \leq k^* \leq n$.

A naive way to determine the jth column of P, P_j, is to let $P_j = e^k$ if $j = k^*$. This certainly ensures that the jth decision will be selected if the environment is e^k. However, this will not work appropriately, since the same decision d_j may be taken for different environment observations. In fact, n is usually much less than N. By using an interval rule matrix like the one constructed in the previous section, one may come to the same decision even with different values of environment observations.

6.3.1 A Straightforward Approach

Our objective is to extract an interval-valued rule matrix $\mathbf{P} = \{\boldsymbol{p}_{ij}\}_{m\times n} = \{[\underline{p}_{ij}, \overline{p}_{ij}]\}_{m\times n}$ from a dataset $\boldsymbol{E} = \{(\boldsymbol{e}^k, d_{k^*})|1 \le k \le N\}$ without any previous knowledge except the training dataset itself. A straightforward approach would be as follows.

For any given $j \in \{1, 2, \ldots, n\}$, we find an interval vector $\boldsymbol{P}_j$ such that $\forall k \in \{1, 2, \ldots, N\}$, $\boldsymbol{e}^k \subset \boldsymbol{P}_j$ if $d_{k^*} = j$. It is ideal if $\boldsymbol{P}_j \cap \boldsymbol{P}_k = \emptyset$ whenever $j \neq k$. Then these mutually exclusive n column interval vectors form an interval rule matrix that fits the training dataset. In the following example, we presorted the training dataset according to the decisions. By taking a hull for each feature corresponding to the same decisions, we easily obtain an interval rule matrix.

Example 2. An environment consists of three features, and there are three possible decisions: a, b, and c. Construct an interval rule matrix from the following collection of desired environment-decision pairs:

```
([0.8, 1.1] [1.1, 1.2] [0.8, 1.7], a);
([0.7, 0.9] [1.3, 1.4] [0.9, 1.1], a);
([0.8, 1.0] [1.3, 1.5] [1.0, 1.8], a);
([0.7, 0.9] [1.2, 1.4] [0.9, 1.0], a);
([0.8, 0.9] [1.2, 1.4] [0.9, 1.0], a).
([0.1, 0.4] [2.0, 2.1] [0.2, 0.5], b);
([0.2, 0.3] [2.0, 2.3] [0.2, 0.4], b);
([0.1, 0.4] [2.0, 2.1] [0.2, 0.5], b);
([0.3, 0.4] [2.1, 2.2] [0.3, 0.4], b);
([0.0, 0.2] [2.2, 2.5] [0.4, 0.5], b);
([1.5, 3.3] [0.7, 0.9] [5.8, 6.5], c);
([1.6, 3.0] [0.7, 1.0] [4.8, 5.0], c);
([2.0, 3.2] [0.9, 1.0] [5.1, 6.3], c);
([3.2, 4.0] [0.4, 1.0] [6.2, 6.4], c);
([1.5, 4.0] [0.5, 1.0] [4.8, 6.0], c);
```

Let us consider the interval vector $\boldsymbol{P}_a$ first. The first element of the interval vector $\boldsymbol{P}_a$ should contain the intervals [0.8, 1.1], [0.7, 0.9], [0.8, 1.0], [0.7, 0.9], and [0.8, 0.9]. Hence, the union of these five intervals, [0.7, 1.1], works. Similarly, we get the second and the third elements of $\boldsymbol{P}_a$ as [1.1, 1.5] and [0.8, 1.8]. Therefore,

$$\boldsymbol{P}_a = \begin{pmatrix} [0.7,\ 1.1] \\ [1.1,\ 1.5] \\ [0.8,\ 1.8] \end{pmatrix}.$$

We can find $\boldsymbol{P}_b$ and $\boldsymbol{P}_c$ similarly and then construct an interval rule matrix as

$$\begin{pmatrix} [0.7, 1.1] & [0.0, 0.4] & [1.5, 4.0] \\ [1.1, 1.5] & [2.0, 2.5] & [0.4, 1.0] \\ [0.8, 1.8] & [0.2, 0.5] & [4.8, 6.5] \end{pmatrix}$$

From the above example we can see that the rule matrix can be updated easily. Whenever a new environment-decision pair becomes available, a union operation on interval vectors takes care of the updating. Let $([0.9, 1.7], [1.2, 1.6], [1.0, 1.5]; a)$ be a newly available pair. Then the first column of the above rule matrix can be adjusted as

$$\begin{pmatrix} [0.7, 1.1] \\ [1.1, 1.5] \\ [0.8, 1.8] \end{pmatrix} \cup \begin{pmatrix} [0.9, 1.7] \\ [1.2, 1.6] \\ [1.0, 1.5] \end{pmatrix} = \begin{pmatrix} [0.7, 1.7] \\ [1.1, 1.6] \\ [0.8, 1.8] \end{pmatrix}.$$

This updating scheme also suggests a way to construct an interval rule matrix starting from an empty rule matrix (i.e., from an initial rule matrix in which every element is an empty interval).

Let us write the straightforward approach as an algorithm that constructs an $m \times n$ interval rule matrix with a known dataset that contains N environment-decision pairs.

Algorithm 3 (Straightforward construction of an interval rule matrix)

- *Initialize an empty* $m \times n$ *rule matrix* $\boldsymbol{P}$ *such that* $\mathbf{p}_{ij} = \emptyset$, $\forall i = 1, 2, \ldots, m$ *and* $j = 1, 2, \ldots, n$.
- *For each environment-decision pair* $(\mathbf{e}^k, d_{k^*})$, *where* $k \in \{1, 2, \ldots, N\}$ *and* $k^* \in \{1, 2, \ldots, n\}$, *update the rule matrix as follows:*

$$\mathbf{p}_{ik^*} \leftarrow hull(\mathbf{p}_{ik^*}, \mathbf{e}_i^k) \quad \text{for } i = 1, 2, \ldots, m$$

where

$$hull([\underline{x}, \overline{x}], [\underline{y}, \overline{y}]) = [\min\{\underline{x}, \underline{y}\}, \max\{\overline{x}, \overline{y}\}]$$

is the smallest interval containing both $\boldsymbol{x} = [\underline{x}, \overline{x}]$ *and* $\boldsymbol{y} = [\underline{y}, \overline{y}]$.

For each environment-decision pair, one column of the rule matrix needs to be updated. Therefore, Algorithm 3 is $\mathcal{O}(mN)$. As we can see in this algorithm, it is unnecessary to presort a training dataset.

The straightforward approach is simple enough, with a relatively low computational cost. However, it may not always result in mutually exclusive columns. We say that an interval rule matrix is "fat" if two or more columns have a nonempty intersection. For example, let us assume that a new data item $([0.3, 0.8][1.4, 2.1][0.4, 0.9]; b)$ becomes available for updating the above rule matrix. It would result in the second column of the rule matrix being

$$\begin{pmatrix} [0.0, 0.4] \\ [2.0, 2.5] \\ [0.2, 0.5] \end{pmatrix} \cup \begin{pmatrix} [0.3, 0.8] \\ [1.4, 2.1] \\ [0.4, 0.9] \end{pmatrix} = \begin{pmatrix} [0.0, 0.8] \\ [1.4, 2.5] \\ [0.2\ 0.9] \end{pmatrix}.$$

The intersection of the updated column with the original first column is nonempty since

$$\begin{pmatrix} [0.7, 1.7] \\ [1.1, 1.6] \\ [0.8, 1.5] \end{pmatrix} \cap \begin{pmatrix} [0.0, 0.8] \\ [1.4, 2.5] \\ [0.2\ 0.9] \end{pmatrix} = \begin{pmatrix} [0.7, 0.8] \\ [1.4, 1.6] \\ [0.8\ 0.9] \end{pmatrix} \neq \emptyset.$$

If an observation falls into the intersection, say

$$\boldsymbol{e} = \begin{pmatrix} [0.72, 0.75] \\ [1.4, 1.5] \\ [0.83\ 0.89] \end{pmatrix},$$

then it is unclear whether one should select a or b as the decision. For this reason, we need to study more sophisticated algorithms.

6.3.2 A Divide-and-Conquer Approach

We now present an alternative approach, with the divide-and-conquer approach, to establish an interval rule matrix $\boldsymbol{P}$ from a given training dataset. In real applications, observed environment features at a particular time are mostly thin intervals. Instead of starting with decisions as in the straightforward approach, let us begin with the consideration of feature parameters.

To avoid "fat" interval rule matrices, we first subdivide the range of each feature parameter into narrow subintervals. Then by picking subintervals from each of the m-features, we construct an m-dimensional tube (interval vectors whose component intervals are narrow). Thus, if the interval for the i-th feature is subdivided into j_i subintervals for each i, $1 \leq i \leq m$, the total number of tubes will be $\prod_{i=1}^{m} j_i$.

For each environment-decision pair in the training dataset, we try to fit the pair with one or more tubes according to its environment value. If a pair can be put inside a tube completely, we increment the frequency of the decision(s) associated with the tube. If the environment of a pair covers several adjacent tubes, we increment the frequency of the decision for each tube that the environment covers. After doing so for all pairs in E, we have a decision frequency list associated with each tube. (Some of the frequencies can be zero.) If these tubes are narrow enough, then each of them may contain only a single decision, say d_j.

In the conquer stage, we select one of the tubes with the highest frequency for d_j as the base for $\boldsymbol{P}_j$, the j-th column of the rule matrix. Adjacent tubes associated with the same decision d_j can then be combined to form the j-th column of the rule matrix. On the other hand, if a tube contains different decisions, it can be subdivided further. Of course, the frequency should be recounted in the latter case.

The conquer stage not only consists of combining those adjacent tubes associated with the same decision but also of dealing with tubes that are not associated with any decisions at all. In addition, a decision may be associated with nonadjacent columns. Those tubes that are not associated with any decisions we call *empty tubes*. These empty tubes can be either removed or kept inactive for possible future use. (Using a predetermined frequency threshold, one may computationally filter out statistically insignificant tubes as empty.)

A decision may be associated with multiple disconnected tubes (i.e., the tubes do not share a common boundary). (Geometrically, there are other tubes between tubes associated with the same decision.) A separation of two disjoint tubes with the same decision is *removable* if the tubes in between are empty. By taking a hull, one may remove such removable separations.

However, there are nonremovable separations, which means that there exists a tube between them but associated with another decision. In such a case, we have to associate the decision with multiple columns. However, this will cause a contradiction with the assumption that the rule matrix has only n columns. To resolve this, we can use multiple rule matrices, such that each rule matrix is on a separate "page." In this way, the rule matrix is no longer a two-dimensional array. The column of a rule matrix can be viewed as a pointer that points to multiple interval vectors on different pages. Obviously, the number of pages should be the same as the maximum number of disjoint tubes that are associated with a single decision.

Let us summarize the above idea in an algorithmic format.

Algorithm 4 (Divide and conquer scheme to construct an interval rule matrix)

1. *Divide*
   a) *Initialization: Subdivide each domain of the m-features into a predetermined number of subintervals to form a set of tubes T. For each $t \in T$, associate a frequency zero with it for each possible decision.*
   b) *For each $(\mathbf{e}^k, d_k^*) \in E$, if $\mathbf{e}^k \cap t \neq \emptyset$ for some $t \in T$ and there are no decisions other than d_k^* associated with t, add 1 to the frequency count of d_k^* on t. Otherwise, if there are decisions other than d_k^* associated with t, then subdivide t for decision separation.*
2. *Conquer*
   a) *For those tubes with the same decision, combine them if they are adjacent or the separations are removable.*
   b) *Form the j-th column of the rule matrix with the tube associated with the decision d_j. Note: There can be multiple separated columns associated with the same decision d_j.*

Notice that the frequency count can be used as an indicator for the strength of a rule. By ignoring very low frequencies, it can be used to avoid tube refinements that are too detailed. To control the output rule matrix with Algorithm 4, one may apply (a) a predefined tube size, (b) a frequency cutoff, and (c) multiple pages. By associating a decision with disconnected columns on multiple pages, the "fat" rule matrix problem in Algorithm 3 may be eliminated.

The computational complexity of Algorithm 4 can be much higher than that of Algorithm 3. For an m-feature environment, if one performs s subdivisions for each feature, then T consists of s^m tubes. Each matching requires $m \log s$ comparisons with binary searching. Even without the tube refinement, the algorithm requires $\mathcal{O}(Nm \log s)$ comparisons. With the refinement, say performing up to h bisections for each feature, the complexity will be $\mathcal{O}(Nmh^m \log s)$. In real applications, the number of features of a complex environment can be very large.

A reasonable approach to make Algorithm 4 more practical is to control the number of features under consideration. Feature selection itself is a computational decision-making problem [7]. To control the number of features, one may rank features according to their correlations with respect to decisions. By selecting subsets of features based on reliably assessing the statistical significance of the relevance of features to a given predictor, one may build more compact feature subsets. Also, applying a singular value decomposition (SVD), one can form a set of features that are linear combinations of the original variables, which provide the best possible reconstruction of the original data in the least squares sense.

In practice, instead of using all of the data in a training dataset to select features, one may first sample the training dataset with Algorithm 4 to form an initial rule matrix. With the initial rule matrix, features can then be selected. Applying Algorithm 4 with these selected features, one can adaptively update the rule matrix. Hopefully, these selected features are good enough for most data, and only a few environment-decision pairs need to have more features than those in the selected sample when updating the rule matrix. Some heuristics such as sample selection, feature granularity, and cutoff threshold are needed in applications.

The above practical approach also has other advantages. It allows one to make a decision with the current rule matrix without completing the training. The correctness of the decision selected can be used as a feedback to adjust the rule matrix. We call this capability the ability to do "online" dynamical training. The selected features can also be updated adaptively. Rules and significance of a feature for decision making can be changed from time to time. An adaptive approach can update them as well. Also, we should point out that domain knowledge can and should play a very significant role in rule matrix generation and feature selection.

6.4 Decision Making with an Interval Rule Matrix

The purpose of establishing an interval rule matrix is to apply it to making decisions. Since a multipage rule matrix can be processed page by page, we consider only single-page interval rule matrices here.

Making a decision based on an environment observation **e** according to an interval rule matrix requires to determine a j such that $\boldsymbol{e}$ "matches" the j-th column of the rule matrix. By the word "match", we mean that the two interval vectors should somehow be "close." To determine the closeness of two interval vectors, we first need to define the distance between two intervals.

Definition 1. *Let $\boldsymbol{a}$ and $\mathbf{b}$ be two intervals. The distance between $\boldsymbol{a}$ and $\mathbf{b}$ is defined as*

$$\operatorname{dist}(\boldsymbol{a},\boldsymbol{b}) = \begin{cases} 0 & \text{if } \boldsymbol{a} \subseteq \boldsymbol{b} \text{ or } \boldsymbol{b} \subseteq \boldsymbol{a} \\ \min\{|a-b|+1;\ \forall a \in \boldsymbol{a}, b \in \boldsymbol{b}\} & \text{if } \boldsymbol{a} \cap \boldsymbol{b} = \emptyset \\ 1 - \dfrac{w(\boldsymbol{a} \cap \boldsymbol{b})}{\min\{w(\boldsymbol{a}), w(\boldsymbol{b})\}} & \text{otherwise.} \end{cases}$$

The above definition implies the following properties:

1. The distance is reflexive: The distance from an interval $\boldsymbol{a}$ to another interval $\boldsymbol{b}$ is the same as from $\boldsymbol{b}$ to $\boldsymbol{a}$.
2. The distance between two intervals is zero if one is a subset of the other.
3. The minimum distance between two disjoint intervals is greater than or equal to 1.
4. The distance between two partially overlapped intervals is between 0 and 1.
5. The distance between two disjoint intervals is in fact the same as

$$|m(\boldsymbol{a}) - m(\boldsymbol{b})| - \frac{w(\boldsymbol{a}) + w(\boldsymbol{b})}{2} + 1,$$

where $m()$ and $w()$ are midpoint and width functions, respectively.

The proof is straightforward.

Note that Definition 1 is appropriate for knowledge processing, but it differs from other commonly used measures of distance, such as the Hausdorff distance [1, p. 11], used within the interval mathematics community.

Applying the concept of distance between intervals, we can define the distance between two interval vectors as follows.

Definition 2. *Let $\boldsymbol{x} = (\boldsymbol{x}_1, \boldsymbol{x}_2, \ldots, \boldsymbol{x}_m)$ and $\boldsymbol{y} = (\boldsymbol{y}_1, \boldsymbol{y}_2, \ldots, \boldsymbol{y}_m)$ be two m-dimensional interval vectors. Then the l-distance between $\boldsymbol{x}$ and $\boldsymbol{y}$ is defined as*

$$\operatorname{dist}_l(\boldsymbol{x},\boldsymbol{y}) = \left(\sum_{1\le i\le m} \operatorname{dist}^l(\boldsymbol{x}_i,\boldsymbol{y}_i) \right)^{\frac{1}{l}}.$$

By computing a distance, say $l = 1$ or $l = \infty$, between an environment observation vector $\boldsymbol{e}$ and each column of an interval rule matrix $\boldsymbol{P}$, one can then make a decision accordingly. It seems reasonable to select the d_j such that $\operatorname{dist}(\boldsymbol{e}, \boldsymbol{P}_j) = \min_{1\le i\le n} \operatorname{dist}(\boldsymbol{e}, \boldsymbol{P}_i)$, where $\boldsymbol{P}_i$ is the i-th column of $\boldsymbol{P}$. This is because the distance between $\boldsymbol{e}$ and $\boldsymbol{P}_i$ is viewed as the strength indicator of the i-th decision. The smaller the distance is, the stronger the decision should be. In other words, the larger the distance is, the less the decision should be selected. Zero distance indicates the strongest match between two interval vectors.

It is possible that an observation vector may have the same (or almost equal) minimum distance to multiple columns of the rule matrix. This can happen even when the observation vector is a thin interval vector. To make a reasonable decision from multiple best matches, instead of a random pick, we define the expected value of a decision as follows.

Definition 3. *Let ρ_j and v_j be the probability and the benefit value of a decision j, respectively. Then, the expected value of the decision j is defined as $exp_j = \rho_j v_j$.*

In constructing an interval rule matrix from a training dataset, the decision frequencies have been recorded. They can be considered as empirical probabilities for each of the decisions. Also, historical data can provide the average return on a decision. Therefore, one may select the decision with the highest expected values among the multimatched columns.

Instead of applying the concept of interval distance, an earlier alternative approach is to select a decision based on an interval rule matrix with fuzzy logic and values of possibility and necessity functions. Readers may refer to [4] and Chapter 2 of this book for more information.

6.5 Conclusions

We have studied an interval rule matrix model for establishing decision-making systems. Using a training dataset consisting of environment-decision pairs, we have proposed two algorithms to generate an interval rule matrix. The straightforward approach has a time complexity of $\mathcal{O}(mN)$. The divide-and-conquer approach may use adaptive modification, "online" training, and feature selection for more practicality.

With an interval rule matrix, making a decision for an environment observation becomes finding the minimum distance between interval vectors. Interval rule matrices have potential applications in rule-based automated decision-making systems.

Acknowledgment: This work is partially supported by the U.S. National Science Foundation under grants CISE/CCF-0202042 and CISE/CCF-0727798.

References

1. Alefeld, G., Herzberger, J.: Introduction to Interval Computations. Academic Press Inc., New York (1983). Translation by J. Rokne from the original German "Einführung In Die Intervallrechnung"
2. Bace, R., Mell, P.: NIST special publication on intrusion detection system. Technical report, NIST (National Institute of Standards and Technology) (2001). http://csrc.nist.gov/publications/nistpubs/800-31/sp800-31.pdf.
3. Berleant, D., Cheong, M.P., Chu, C.C.N., Guan, Y., Kamal, A., Sheble, G., Ferson, S., Peters, J.F.: Dependable handling of uncertainty. Reliable Computing 9(6), 407–418 (2003)
4. de Korvin, A., Hu, C., Chen, P.: Generating and applying rules for interval valued fuzzy observations. In: Z.R. Yang, R.M. Everson, H. Yin (eds.) Intelligent Data Engineering and Automated Learning, Lecture Notes in Computer Science, Vol. 3177, pp. 279–284. Springer-Verlag, Heidelberg (2004)
5. de Korvin, A., Hu, C., Sirisaengtaksin, O.: On firing rules of fuzzy sets of type II. Applied Mathematics 3, 151-159 (2000)
6. Duan, Q., Hu, C., Wei, H.C.: Enhancing network intrusion detection systems with interval methods. In: Proceedings of the ACM Symposium on Applied Computing, pp. 1444–1448 (2005)
7. Guyon, I., Elisseeff, A.: An introduction to variable and feature selection. J. Machine Learning Research 3, 1157–1182 (2003)
8. Ilgun, K., Kemmerer, R.A., Porras, P.A.: State transition analysis: A rule-based intrusion detection approach. Software Engineering 21(3), 181–199 (1995)
9. Jang, J.S.R.: ANFIS: Adaptive-network-based fuzzy inference system. IEEE Transactions on Systems, Man, and Cybernetics 23, 665–684 (1993)
10. Julisch, K.: Clustering intrusion detection alarms to support root cause analysis. ACM Transactions on Information and System Securroty 6(4), 443–471 (2003)
11. Mendel, J.: Uncertain Rule-Based Fuzzy Logic Systems: Introduction and New Directions. Prentice-Hall, Upper Saddle River, NJ (2001)
12. Pawlak, Z.: Rough sets and fuzzy sets. Fuzzy Sets and Systems 17, 99–102 (1985)
13. Seleznyov, A., Puuronen, S.: Anomaly intrusion detection systems: Handling temporal relations between events. In: Proceedings of Recent Advances in Intrusion Detection, West Lafayette, IN (1999). http://www.raid-symposium.org/raid99/PAPERS/Seleznyov.pdf
14. Shary, S.P.: A new technique in systems analysis under interval uncertainty and ambiguity. Reliable Computing 8(5), 321–418 (2002)
15. Zadeh, L.A.: Outline of a new approach to the analysis of complex systems and decision processes. IEEE Transactions on Systems, Man, and Cybernetics SMC-3, 28–44 (1973)

7

Interval Matrix Games

W. Dwayne Collins[1] and Chenyi Hu[2]

[1] Department of Mathematics and Computer Science, Hendrix College, 1600 Washington Avenue, Conway, AR 72032, USA. collins@hendrix.edu
[2] Department of Computer Science, University of Central Arkansas, 201 Donaghey Avenue, Conway, AR 72035-0001, USA. chu@uca.edu

Matrix games have been widely used in decision-making systems. In practice, for the same strategies players take, the corresponding payoffs may be within certain ranges rather than exact values. To model such uncertainty in matrix games, we consider interval-valued game matrices in this chapter and extend the results of classical strictly determined matrix games to fuzzily determined interval matrix games. Finally, we give an initial investigation into mixed strategies for such games. We reported this work initially at the Forging New Frontiers at the University of California, Berkeley in November 2005. The full paper [2] then appeared in Springer's journal Soft Computing in 2008.

7.1 Introduction

7.1.1 Matrix Games

Game theory had its beginnings in the 1920s and significantly advanced at Princeton University through the work of John Nash [3, 7, 8, 10]. The simplest game is a zero-sum game involving only two players. An $m \times n$ matrix $G = \{g_{ij}\}_{m \times n}$ may be used to model such a two-person zero-sum game. If the row player R uses his i-th strategy (row) and the column player C selects her j-th choice (column), then R wins (and subsequently C loses) the amount g_{ij}. The objective of R is to maximize his gain while C tries to minimize her loss.

Example 1. A game is described by the matrix

$$G = \begin{bmatrix} 0 & 6 & -2 & -4 \\ 5 & 2 & 1 & 3 \\ -8 & -1 & 0 & 20 \end{bmatrix}. \tag{7.1}$$

In the above game, the players R and C have three and four possible strategies, respectively. If R chooses his first strategy and C chooses her second, then R

C. Hu et al. (eds.), *Knowledge Processing with Interval and Soft Computing*,
DOI: 10.1007/978-1-84800-326-2_7, © Springer-Verlag London Limited 2008

wins $g_{12} = 6$ (C loses 6). If R chooses his third strategy and C chooses her first, then R wins $g_{31} = -8$ (R loses 8, C wins 8). In this chapter we restrict our attention to such two-person zero-sum games.

7.1.2 Strictly Determined Matrix Games

If there exists a g_{ij} in a classical $m \times n$ game matrix G such that g_{ij} is simultaneously the minimum value of the i-th row and the maximum value of the j-th column of G, then g_{ij} is called a *saddle value* of the game. If a matrix game has a saddle value, it is said to be *strictly determined.* It is well known, [3] and [10], that the optimal strategies for both R and C in a strictly determined game are as follows:

- R should choose any row containing a saddle value.
- C should choose any column containing a saddle value.

A saddle value is also called the value of the (strictly determined) game. In the above example, g_{23} is simultaneously the minimum of the second row and the maximum of the third column. Hence, the game is strictly determined and its value is $g_{23} = 1$. The knowledge of an opponent's move provides no advantage since the optimal strategies for both players will always result in a saddle value as the payoff in a strictly determined game.

7.1.3 Motivation for This Work

Matrix games have many useful applications, especially in decision-making systems. However, in real-world applications, due to certain forms of uncertainty, outcomes of a matrix game may not be a fixed number, even though the players do not change their strategies. Hence, fuzzy games have been studied [4, 9, 11]. By noticing the fact that the payoffs may only vary within a designated range for fixed strategies, we propose using an interval-valued matrix, whose entries are closed intervals, to model this kind of uncertainty.

In this chapter, as throughout this book, we use boldface letters to denote (closed and bounded) intervals. For example, $\boldsymbol{x}$ is an interval. Its greatest lower bound and the least upper bound are denoted by $\underline{x}$ and $\overline{x}$, respectively. We use uppercase letters to denote general matrices. Boldface uppercase letters will represent a interval-valued matrices.

Throughout this chapter, we assume that the intervals in the game matrix $\boldsymbol{G}$ are closed and bounded intervals of real numbers and, for this investigation, represent uniformly distributed possible payoffs.

Definition 1. *Let $\boldsymbol{G} = \{\boldsymbol{g}_{ij}\}$ be an $m \times n$ interval-valued matrix. The matrix $\boldsymbol{G}$ defines a* zero-sum interval matrix game *provided whenever the row player R uses his i-th strategy and the column player C selects her j-th strategy, then R wins and C correspondingly loses a common $x \in \boldsymbol{g}_{ij}$.*

Example 2. Consider the following interval game matrix:

$$G = \begin{bmatrix} [0, 1] & [6, 7] & [-2, 0] & [-4, -2] \\ [5, 6] & [2, 7] & [1, 3] & [3, 3] \\ [-8, -5] & [-1, 0] & [0, 0] & [20, 25] \end{bmatrix} \tag{7.2}$$

In this game, if R chooses row one and C selects column two, then R wins an amount $x \in [6, 7]$. (C loses the same x that R wins.)

In this chapter, we extend results of classical matrix games to interval-valued games. To accomplish this, we need to define fuzzy relational operators for intervals in order to compare every pair of possible interval payoffs from a rational game-play perspective. These relational operators for intervals will be developed in Section 7.2. We then study crisply determined and fuzzily determined interval games in Sections 7.3 and 7.4. Since not all interval games are determined, we begin an investigation of mixed strategies for non-determined games. We describe a potential mapping of such an interval game into an interval linear programming problem in Section 7.5, and we show how linear interval inequalities can be solved under our definition in Section 7.6. We summarize these results in Section 7.7.

7.2 Comparing Intervals

To compare strategies and payoffs for an interval game matrix, we need a notion of an interval ordering relation that corresponds to the intuitive notion of a "better possible" outcome or payoff. This will be done by defining the notion of a nonempty interval $\boldsymbol{x}$ not being a better payoff than a nonempty interval $\boldsymbol{y}$ (i.e., the notion that $\boldsymbol{x}$ is less than or equal to $\boldsymbol{y}$). Other approaches that define such relational orderings between some pairs of intervals have been developed and extended. In [5], Fishburn defined a concept of interval order corresponding to a special kind of partially ordered set. His context is for the study of the order of vertices in interval graphs. An interval graph refers to a graph $(X, \sim)$ whose points can be mapped into intervals of a linearly ordered set such that, for all distinct x and y, $x \sim$ y if and only if the intervals assigned to x and y have a nonempty intersection. Allen's [1] in 1983 listed 13 possible cases for the temporal relationships between two time intervals. However, neither of these two developments compares general intervals or models such a comparison in our game-theoretic context. Unlike these models, we wish to make every pair of our intervals comparable and to fuzzily quantify the notion of "indifference" in our game-theoretic context except when the two intervals are equal.

For the development of our relational operators in our context, we assume that a rational player will not prefer an interval $\boldsymbol{x}$ as in Figure 7.1, Case 1, to interval $\boldsymbol{y}$, as every possible payoff value $x \in \boldsymbol{x}$ is less than every payoff value $y \in \boldsymbol{y}$. Similarly, we assume that in the case of the intervals in Figure 7.1,

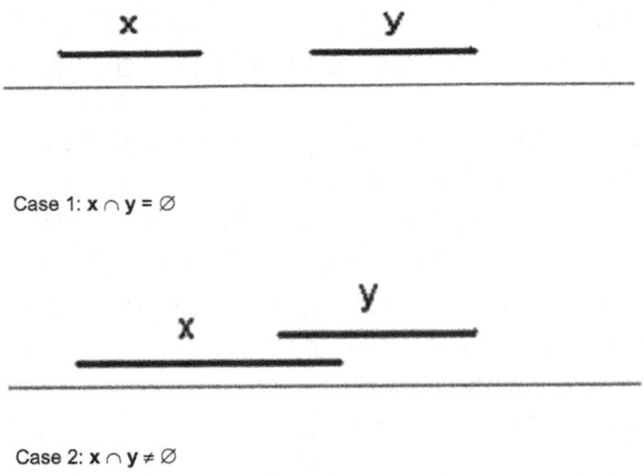

Fig. 7.1. Nonoverlapping and overlapping intervals.

Case 2, the player will not prefer interval $\boldsymbol{x}$ over $\boldsymbol{y}$, since no value in $\boldsymbol{x}$ offers a payoff that is greater than what is possible in $\boldsymbol{y}$, and $\boldsymbol{y}$ offers no payoff that is less than what is possible in $\boldsymbol{x}$. Thus, choosing interval $\boldsymbol{y}$ over $\boldsymbol{x}$ maximizes both the least possible and greatest possible payoff. Finally, in case $\boldsymbol{x} = \boldsymbol{y}$, we assume that a rational player will prefer neither over the other. Therefore, in these cases, using $\leq$ to represent the relation "is not preferred to," we have $\boldsymbol{x} \leq \boldsymbol{y}$ in the cases represented by Cases 1 and 2 and each of $\boldsymbol{x} \leq \boldsymbol{y}$ and $\boldsymbol{y} \leq \boldsymbol{x}$ when $\boldsymbol{x}$ is equal to $\boldsymbol{y}$. In these cases, the preference order exhibits the properties of a total order. Hence, these comparisons can be crisply defined as true and are consistent with traditional interval comparison operators.

When $\boldsymbol{x}$ is completely contained in $\boldsymbol{y}$, as displayed in Figure 7.2, the notion of payoff preference becomes uncertain, since there exist payoff values in $\boldsymbol{y}$ that are less than every possible payoff in $\boldsymbol{x}$ as well as values in $\boldsymbol{y}$ that are greater than every possible payoff in $\boldsymbol{x}$. In this case, a risk-adverse player may (but not necessarily will) prefer $\boldsymbol{x}$ to $\boldsymbol{y}$, since $\boldsymbol{x}$ contains the largest worst possible actual payoff value, whereas a (rational) risk-taking player may prefer $\boldsymbol{y}$ to $\boldsymbol{x}$, since $\boldsymbol{y}$ contains the largest best possible actual payoff. However, for any single game, either player may also rationally decide that he/she is indifferent to the two choices or will choose the other. In other words, the interval payoff preference cannot be determined with classical binary logic. This uncertainty, however, can be well addressed with the theory of fuzzy logic developed by Zadeh [12]. Therefore, we extend the previous crisp preference comparisons with fuzzy membership. Such a fuzzy membership extension might be expected to be a continuous one in terms of holding one interval fixed and moving the other in terms of its midpoint and width, but in the presented context, no such

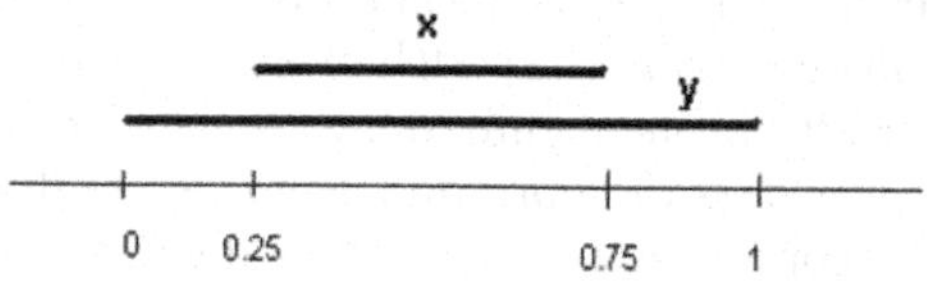

Fig. 7.2. Nested intervals.

continuous extension is possible. To see this, observe that if the widths of $\boldsymbol{x}$ and $\boldsymbol{y}$ are equal and the two intervals are initially positioned as in Case 1 of Figure 7.1, as $\boldsymbol{x}$ moves to the right, the inequality $\boldsymbol{x} \leq \boldsymbol{y}$ is crisply true (having membership value 1 in a fuzzy context) until $\boldsymbol{x} = \boldsymbol{y}$ and is crisply false (having membership value 0 in a fuzzy context) afterward. Hence, no membership value of "$\boldsymbol{x}$ is not preferred to $\boldsymbol{y}$" will allow for a continuous extension.

To fuzzily quantify uncertainty as in Figure 7.2, we consider the case that the interval $\boldsymbol{x}$ is positioned with its left endpoint the same as the left endpoint of $\boldsymbol{y}$ and $\boldsymbol{x} \subset \boldsymbol{y}$. In this case, a rational player will crisply prefer $\boldsymbol{y}$ over $\boldsymbol{x}$ for the same reasons expressed in the analysis of Figure 7.1. Hence, $\boldsymbol{x} \leq \boldsymbol{y}$ crisply, and in terms of a fuzzy relational operator, the membership value of this relation is 1. On the other hand, when $\boldsymbol{x}$ is positioned to share its right endpoint with $\boldsymbol{y}$, a rational player will crisply prefer $\boldsymbol{x}$ to $\boldsymbol{y}$ for the same reason. Hence, in this case the membership value of $\boldsymbol{x} \leq \boldsymbol{y}$ is 0. We then define the fuzzy membership to be a linear mapping from 1 to 0 as the interval $\boldsymbol{x}$ "moves" from right to left. The corresponding fuzzy membership values of this relation then can be associated with the notion of the degree of risk-taking that a player may exhibit. However, this relationship is not a probabilistic one, but rather a possible one. For example, a risk-adverse player facing a choice between two such intervals with an $\boldsymbol{x} \leq \boldsymbol{y}$ membership value close to 1 may consider the risk of choosing $\boldsymbol{y}$ over $\boldsymbol{x}$, in spite of the possibility of receiving an actual payoff less than every value in $\boldsymbol{x}$. On the other hand, a risk-taking player may choose $\boldsymbol{y}$ over $\boldsymbol{x}$ with a small positive membership value of $\boldsymbol{x} \leq \boldsymbol{y}$.

The linear map[3]

$$f(\boldsymbol{x}, \boldsymbol{y}) = \frac{\overline{y} - \overline{x}}{w(y) - w(x)} \tag{7.3}$$

meets the requirement, where $w(x) = \overline{x} - \underline{x}$ is the width of the interval $\boldsymbol{x}$.

As a special instance, note that the membership is 0.5 when the midpoints of $\boldsymbol{x}$ and $\boldsymbol{y}$ coincide. If one keeps the interval $\boldsymbol{y}$ fixed, one keeps the midpoints of $\boldsymbol{x}$ and $\boldsymbol{y}$ equal, and one allows the width of $\boldsymbol{x}$ to vary continuously, there is a pronounced discontinuity in the membership values of $\boldsymbol{x} \leq \boldsymbol{y}$ when the widths become equal. However, this discontinuity is not in conflict with the measure of uncertainty of the comparison, since by our definition there is uncertainty in the comparison at all widths of $\boldsymbol{x}$ except when the intervals are equal.

Summarizing the above discussion, we extend the crisp comparison operator by defining the fuzzy comparison operator $\preceq$ for two closed and bounded intervals for the "not preferred to" relationship as follows.

Definition 2. *Let $\boldsymbol{x}$ and $\boldsymbol{y}$ be two nontrivial intervals. The binary fuzzy operator $\preceq$ of $\boldsymbol{x}$ and $\boldsymbol{y}$ returns the membership for "$\boldsymbol{x}$ is not preferred to $\boldsymbol{y}$" between 0 and 1 as*

$$\boldsymbol{x} \preceq \boldsymbol{y} = \begin{cases} 1 & \underline{x} \leq \underline{y} \leq \overline{x} < \overline{y} \\ \dfrac{\overline{y} - \overline{x}}{w(\boldsymbol{y}) - w(\boldsymbol{x})} & \underline{y} < \underline{x} < \overline{x} \leq \overline{y}, w(\boldsymbol{x}) \neq w(\boldsymbol{y}) \\ 1 & \underline{x} = \underline{y}, w(\boldsymbol{x}) = w(\boldsymbol{y}) \\ 0 & \textit{otherwise}. \end{cases} \tag{7.4}$$

One can define the dual fuzzy relation "is preferred to" in the analogous way. We will use the symbol $\succeq$ to denote this dual relationship as a reminder of the antisymmetry in the crisp case. Therefore, $\succeq$ can be defined in terms of $\preceq$ as follows.

Definition 3. *The binary fuzzy operator $\succeq$ of two intervals $\boldsymbol{x}$ and $\boldsymbol{y}$ is defined as $\boldsymbol{x} \succeq \boldsymbol{y} = 1$ if $\boldsymbol{x} = \boldsymbol{y}$, and $\boldsymbol{x} \succeq \boldsymbol{y} = 1 - (\boldsymbol{x} \preceq \boldsymbol{y})$ otherwise.*

Definition 4. *If the value of $\boldsymbol{x} \preceq \boldsymbol{y}$ is exactly 1 or 0, then we say that $\boldsymbol{x}$ and $\boldsymbol{y}$ are crisply comparable . Otherwise, we say that they are fuzzily comparable.*

7.3 Crisply Determined Interval Matrix Games

In this section, we extend the concept of classical strictly determined games to interval matrix games whose row and column entries are crisply comparable. In this case, we will use $\leq$ and $\geq$ in place of $\succeq$ and $\preceq$ to emphasize the crispness of the appropriate interval comparisons.

[3] Linear in the position of $\overline{\boldsymbol{x}}$ as $\boldsymbol{y}$ is held fixed and the width of $\boldsymbol{x}$ is held fixed.

Definition 5. *Let $\boldsymbol{G}$ be a $m \times n$ interval game matrix. If there exists a $\boldsymbol{g}_{ij} \in \boldsymbol{G}$ such that $\boldsymbol{g}_{ij}$ is simultaneously crisply less than or equal to $\boldsymbol{g}_{ik}$ for all $k \in \{1, 2, \ldots, n\}$ and crisply greater than or equal to $\boldsymbol{g}_{lj}$ for all $l \in \{1, 2, \ldots, m\}$ then the interval $\boldsymbol{g}_{ij}$ is called a saddle interval of the game. An interval matrix game is crisply determined if it has a saddle interval.*

By Definition 5, to determine whether an interval game matrix is crisply determined, one needs only to do the following:

1. For each row ($1 \leq i \leq m$), find an entry $\boldsymbol{g}_{ij^*}$ that is crisply less than or equal to all other entries in the i-th row.
2. For each column ($1 \leq j \leq n$), find an entry $\boldsymbol{g}_{i^*j}$ that is crisply greater than or equal to all other entries in the j-th column.
3. Determine if there is an entry $\boldsymbol{g}_{i^*j^*}$ that is simultaneously a minimum of the i-th row and a maximum of the j-th column.
4. If any of the above values cannot be found, the game is not crisply determined. Otherwise, it is a crisply determined interval matrix game.

Example 3. Examining the interval game matrix (7.2), we found that $\boldsymbol{g}_{14}$, $\boldsymbol{g}_{23}$, and $\boldsymbol{g}_{31}$ are the minima of rows 1, 2, and 3, respectively. Similarly, $\boldsymbol{g}_{21}$, $\boldsymbol{g}_{12}$, $\boldsymbol{g}_{23}$, and $\boldsymbol{g}_{34}$ are the maxima of columns 1, 2, 3, and 4, respectively. Furthermore, $\boldsymbol{g}_{23}$ is simultaneously the minimum of the second row and the maximum of the third column. Hence, $\boldsymbol{g}_{23} = [1, 3]$ is a saddle interval of the game matrix. This is a crisply determined interval matrix game.

Mimicking the optimal strategy for a classical strictly determined game, we have the optimum strategies for both R and C in a crisply determined interval matrix game defined as follows:

- R should choose any row containing a saddle interval.
- C should choose any column containing a saddle interval.

In this case, uniqueness of the saddle interval value can be established.

Theorem 1. *If an interval matrix game is crisply determined, its saddle intervals are identical.*

Proof. Let $\boldsymbol{G}$ be a crisply determined interval game matrix and $\boldsymbol{g}_{ij}$ and $\boldsymbol{g}_{lk}$ are saddle intervals. Then $\boldsymbol{g}_{ij} \leq \boldsymbol{g}_{ik} \leq \boldsymbol{g}_{lk}$ and $\boldsymbol{g}_{ij} \geq \boldsymbol{g}_{lj} \geq \boldsymbol{g}_{lk}$. Hence, from Definitions 2 and 3, $\boldsymbol{g}_{ij} = \boldsymbol{g}_{lk}$.

As in the classical case, in a strictly determined interval game, the knowledge of an opponent's move provides no advantage, since the payoff is assumed to be uniformly distributed within a saddle interval.

Definition 6. *The* value interval *of a strictly determined interval game is its saddle interval. A strictly determined interval game is fair if its saddle interval is symmetric with respect to zero (i.e., if the saddle interval is of the form $[-a, a]$ for $a \geq 0$). A strictly determined interval game that is not fair is said to be unfair.*

From Example 3 we know that $\boldsymbol{g}_{23}$ is a saddle interval of the matrix game (7.2). However, the midpoint of $\boldsymbol{g}_{23}$ is 2. Hence, the game is unfair, since the row player has an average advantage of 2.

7.4 Fuzzily Determined Interval Matrix Games

For a general interval game matrix, crisp comparability may not be satisfied for all intervals in the same row (or column). Hence, we now must extend interval comparability to define the fuzzy memberships of an interval $\boldsymbol{v}_i$ being a minimum and a maximum of an interval vector $\boldsymbol{V}$; then we define the notion of a least and greatest interval in $\boldsymbol{V}$.

Definition 7. *Let $\boldsymbol{V} = \{\boldsymbol{v}_1, \boldsymbol{v}_2, \ldots, \boldsymbol{v}_n\}$ be an interval vector. The fuzzy membership of $\boldsymbol{v}_i$ being a least interval in $\boldsymbol{V}$ is defined as*

$$\mu(\boldsymbol{v}_i) = \min_{1 \leq j \leq n} \{\boldsymbol{v}_i \prec \boldsymbol{v}_j\}$$

and a least interval of the vector $\boldsymbol{V}$ is defined as an interval whose μ value is largest, that is, an interval $\boldsymbol{v}_{i^*}$ such that

$$\boldsymbol{v}_{i^*} = \max_{1 \leq i \leq n} \mu(\boldsymbol{v}_i).$$

Likewise, the fuzzy membership of $\boldsymbol{v}_i$ being a maximum interval in $\boldsymbol{V}$ is

$$\nu(\boldsymbol{v}_i) = \min_{1 \leq j \leq n} \{\boldsymbol{v}_i \succeq \boldsymbol{v}_j\}$$

and a greatest interval of the vector $\boldsymbol{V}$ is

$$\boldsymbol{v}_{i^*} = \max_{1 \leq i \leq n} \nu(\boldsymbol{v}_i).$$

Example 4. Find the least and the greatest intervals for the interval vector $\boldsymbol{V} = \{[2,5],[3,7],[4,5]\}$.

Solution: We notice that $\boldsymbol{v}_2$ and $\boldsymbol{v}_3$ are not crisply comparable. By Definition 7, we have $\mu([2,5]) = 1$, $\nu([2,5]) = 0$; $\mu([3,7]) = 0$, $\nu([3,7]) = \frac{2}{3}$; and $\mu([4,5]) = 0$, $\nu([4,5]) = \frac{1}{3}$. Hence, the least interval of the vector $\boldsymbol{V}$ is $\boldsymbol{v}_1 = [2,5]$ with membership 1 and the greatest interval of $\boldsymbol{V}$ is $\boldsymbol{v}_2 = [3,7]$ with membership $\frac{2}{3}$.

Notice, however, that unlike real-valued games, the least or greatest interval of a vector is not necessarily unique. Uniqueness can happen only when unequal intervals share the same midpoint, as the next example shows.

Example 5. Given the interval vector $\boldsymbol{V} = \{[2,5],\ [3,6],\ [4,5]\}$, we find that the least interval of the vector $\boldsymbol{V}$ is $\boldsymbol{v}_1 = [2,5]$ with membership 1. However, as $\nu([2,5]) = 0$, $\nu([3,6]) = \frac{1}{2}$, and $\nu([4,5]) = \frac{1}{2}$, each of $[3,6]$ and $[4,5]$ is a greatest interval with membership value $\frac{1}{2}$.

Definition 7 provides us a way to fuzzily determine least and greatest intervals for any interval vectors. We are now able to define fuzzily determined interval matrix games as follows.

Definition 8. *Let $\boldsymbol{G}$ be an $m \times n$ interval game matrix. If there is a $\boldsymbol{g}_{ij} \in \boldsymbol{G}$ such that $\boldsymbol{g}_{ij}$ is simultaneously a least and a greatest interval for the i-th row and the j-th column of $\boldsymbol{G}$, respectively, then $\boldsymbol{G}$ is a* fuzzily determined interval game. *We also call such $\boldsymbol{g}_{ij}$ a fuzzy saddle interval of the game with its membership as* $\min\{\mu(\boldsymbol{g}_{ij}), \nu(\boldsymbol{g}_{ij})\}$.

It is obvious that the crisply determined interval game defined in Definition 5 is just a special case of a fuzzily determined interval game with 1 as its membership. The game value of a fuzzily determined interval game can be reasonably defined as its fuzzy saddle interval with the largest membership value.

For the convenience of computer implementations, we summarize our discussion as the following algorithm.

Algorithm 5 (Determine if an interval matrix game is fuzzily determined, and, if so, determine the fuzzy saddle intervals.)

1. *Initialization:*
   a) *Input interval game matrix* $\boldsymbol{G} = \{\boldsymbol{g}_{ij}\}_{m \times n}$.
   b) *Initialize* `FuzzilyDetermined` *to be* `false`.
2. *Calculation:*
   a) *Evaluate* $\mu(\boldsymbol{g}_{ij})$ *and* $\nu(\boldsymbol{g}_{ij})$ *for all* $i = 1$ *to* m *and* $j = 1$ *to* n.
   b) *For each of* $i = 1$ *to* m, *find* j^* *such that* $\mu(\boldsymbol{g}_{ij^*}) = \max_{1 \le j \le n} \{\mu(\boldsymbol{g}_{ij})\}$.
   Note: j^* *depends on* i.
   c) *For each of* $j = 1$ *to* n, *find* i^* *such that* $\nu(\boldsymbol{g}_{i^*j}) = \max_{1 \le i \le m} \{\nu(\boldsymbol{g}_{ij})\}$.
   Note: i^* *depends on* j.
3. *Checking: For each of* $i = 1$ *to* m *and corresponding* j^*, *check if* $\boldsymbol{g}_{ij^*}$ *is also a greatest interval for the* j^* *column. If so:*
   a) *Update* `FuzzilyDetermined` *to* `true`.
   b) *Record* $\boldsymbol{g}_{ij^*}$ *as a fuzzy saddle interval with its membership* $\min\{\mu(\boldsymbol{g}_{ij^*}), \nu(\boldsymbol{g}_{ij^*})\}$.
4. *Finding results:*
   a) *If* `FuzzilyDetermined` *is* `false`, *the interval game is not fuzzily determined.*
   b) *Otherwise, the interval game is fuzzily determined; return the fuzzy saddle intervals that have the largest membership among all recorded fuzzy saddle intervals. Note: The game is crisply determined if the resulting membership is 1.*

The concept of a fuzzily determined interval game in Definition 8 can be further generalized. For each $\boldsymbol{g}_{ij} \in \boldsymbol{G}$, the membership of $\boldsymbol{g}_{ij}$ being simultaneously a least and a greatest interval for the i-th row and the j-th column of

G can be defined as $\varphi(\boldsymbol{g}_{ij}) = \min\{\mu(\boldsymbol{g}_{ij}), \nu(\boldsymbol{g}_{ij})\}$. The entries of $\boldsymbol{G}$ with the largest value of φ can be considered to be fuzzy saddle intervals. Therefore, for any interval game matrix, one can find its fuzzy saddle intervals as those intervals with the largest value of φ. However, it may not make any practical sense if the membership value is too small.

There are many applications of classical game theory to problems in decision theory and finance. In particular, the following is an example of how interval Nash games may apply to determine optimal investment strategies.

Example 6. Consider the case of an investor making a decision on to how to invest a nondivisible sum of money when the economic environment may be categorized into a finite number of states. There is no guarantee that any single value (return on the investment) can adequately model the payoff for any one of the economic states. Hence, it is more realistic to assume that each payoff lies in some interval.

For this example it is assumed that the decision of such an investor can be modeled under the assumption that the economic environment (or nature) is, in fact, a rational "player" that will choose an optimal strategy. Suppose that the options for this player are the following: strong economic growth, moderate economic growth, no growth or shrinkage, and moderate shrinkage (negative growth). For the investor player the options are the following: invest in bonds, invest in stocks, and invest in a guaranteed fixed return account. In this case, clearly a single value for the payoff of either investment in bonds or stock cannot be realistically modeled by a single value representing the percent of return. Hence, a game matrix with interval payoff values better represents the view of the game from both players' perspectives.

Consider the following interval game matrix for this scenario, where the percentage of return is represented in decimal form:

	Bonds	Stocks	Fixed
Strong	[0.11, 0.136]	[0.125, 0.158]	[0.045, 0.045]
Moderate	[0.083, 0.122]	[0.08, 0.11]	[0.045, 0.045]
None	[0.049, 0.062]	[0.02, 0.042]	[0.045, 0.045]
Negative	[0.022, 0.03]	[−0.04, 0.015]	[0.045, 0.045]

The intervals in each row and column are strictly comparable to each other, and using the techniques described earlier, one finds that the game is strictly determined, with the value of the game the trivial interval $[0.045, 0.045]$. This corresponds to the actions of those investors who do not have any insight into what the economy may do in a given time period and who cannot take high risks.

7.5 Toward Optimal Mixed Strategies Through Linear Programming

As in the case of classical matrix games, there is no guarantee that an interval-valued matrix game is crisply or fuzzily determined. For a nondetermined interval matrix game, one needs to find an optimal mixed strategy for each player. For such nondetermined interval-valued matrix games, we will assume that these mixed strategies are represented by crisp probability values, whose sum for each player is exactly equal to 1. Hence, the goal is to describe a context in which each player can choose an optimal mixed strategy from the set of all possible mixed strategies.

We first remind the reader of the traditional meaning of mixed strategy.

Definition 9. *Suppose* $\boldsymbol{G}$ *is an* $m \times n$ *matrix game (interval or otherwise). Then a* mixed strategy *for the row player is a set of probabilities* $P = (p_1, p_2, \ldots, p_m)$*, such that the player selects row* i *with probability* p_i*. Similarly, a mixed strategy for the column player is a set of probabilities* $(q_1, \ldots, q_n)$*, such that the column player selects the* j*-th column with probability* q_j*.*

In the classical zero-sum matrix game context, the problem of finding an optimal mixed strategy solution can be mapped to an equivalent linear programming problem. We will now investigate such a transformation for interval-valued games and present the resulting linear programming problems to be solved.

Suppose $\boldsymbol{G} = (\boldsymbol{g}_{ij})$ is an $m \times n$ interval game matrix and the column player C chooses column j as her strategy. If $P = (p_1, p_2, \ldots, p_m)$ is the row player's mixed strategy, then the *expected value* for the row player, given C's given strategy, is the interval $\boldsymbol{v}$ defined by

$$\boldsymbol{v} = p_1 \cdot \boldsymbol{g}_{1j} + p_2 \cdot \boldsymbol{g}_{2j} + \cdots + p_m \cdot \boldsymbol{g}_{mj} = \sum_{i=1}^{m} p_i \cdot \boldsymbol{g}_{ij}.$$

To find the row player's optimal strategy, we use the "max-min" principle of traditional zero-sum matrix games, namely to find the largest minimum expected value/payoff. Hence, we need to find a "maximum" value $\boldsymbol{v}$ and the corresponding mixed strategy P so that $p_1 \cdot \boldsymbol{g}_{1j} + p_2 \cdot \boldsymbol{g}_{2j} + \cdots + p_m \cdot \boldsymbol{g}_{mj} \succeq \boldsymbol{v}$ for each $1 \leq j \leq n$. The corresponding system to solve is

$$\left\{\begin{array}{c} \text{Maximize } \boldsymbol{v} \text{ subject to} \\ \\ x_1 \cdot \boldsymbol{g}_{11} + x_2 \cdot \boldsymbol{g}_{21} + \cdots + x_m \cdot \boldsymbol{g}_{m1} \succeq \boldsymbol{v} \\ x_1 \cdot \boldsymbol{g}_{12} + x_2 \cdot \boldsymbol{g}_{22} + \cdots + x_m \cdot \boldsymbol{g}_{m2} \succeq \boldsymbol{v} \\ \vdots \\ x_1 \cdot \boldsymbol{g}_{1n} + x_2 \cdot \boldsymbol{g}_{2n} + \cdots + x_m \cdot \boldsymbol{g}_{mn} \succeq \boldsymbol{v} \\ \sum_{i=1}^{m} x_i = 1 \\ x_1, x_2, \cdots, x_m \geq 0. \end{array}\right\} \tag{7.5}$$

Since the entries of the game matrix $\boldsymbol{G}$ represents the gains to the row player, the column player attempts to minimize her losses. Therefore, she attempts to find the smallest maximum expected value, and the corresponding (dual) system for her is

$$\left\{\begin{array}{c} \text{Minimize } \boldsymbol{v} \text{ subject to} \\ \\ x_1 \cdot \boldsymbol{g}_{11} + x_2 \cdot \boldsymbol{g}_{12} + \cdots + x_n \cdot \boldsymbol{g}_{1n} \preceq \boldsymbol{v} \\ x_1 \cdot \boldsymbol{g}_{21} + x_2 \cdot \boldsymbol{g}_{22} + \cdots + x_n \cdot \boldsymbol{g}_{2n} \preceq \boldsymbol{v} \\ \vdots \\ x_1 \cdot \boldsymbol{g}_{m1} + x_2 \cdot \boldsymbol{g}_{m2} + \cdots + x_n \cdot \boldsymbol{g}_{mn} \preceq \boldsymbol{v} \\ \sum_{i=1}^{n} x_i = 1 \\ x_1, x_2, \cdots, x_m \geq 0 \end{array}\right\} \tag{7.6}$$

In the classical game theory context, one can assume that each of the payoffs is positive, since an appropriate linear shift of the payoff values does not affect the characteristics of the game. In the case of interval-valued games, a similar shift to make each of the interval payoffs positive (i.e., to make the left endpoint of each interval entry in the game matrix positive) can be employed. This shift, as will be shown, does not affect the characteristics of the game.

Theorem 2. *Suppose $\boldsymbol{G} = (\boldsymbol{g}_{ij})$ is an $m \times n$ interval game matrix and $c > 0$. The interval $\boldsymbol{v}$ is a row player's optimal mixed strategy expected value with strategy distribution $P = (p_1, p_2, \ldots, p_m)$ if and only if $\boldsymbol{v} + [c, c]$ is a corresponding optimal value with strategy distribution P for the row player in the game $\boldsymbol{G}' = (\boldsymbol{g}_{ij} + [c, c])$.*

Proof. If $(p_1, p_2, \ldots, p_m)$ is a strategy distribution and $1 \leq j \leq n$, then since each x_i is a real number, and the shift $[c, c]$ is a real number, we have

$$\sum_{i=1}^{m} x_i(\boldsymbol{g}_{ij} + [c, c]) = \sum_{i=1}^{m} (x_i \cdot \boldsymbol{g}_{ij} + x_i \cdot [c, c]) = \sum_{i=1}^{m} x_i \boldsymbol{g}_{ij} + [c, c] \sum_{i=1}^{m} x_i$$

$$= \sum_{i=1}^{m} x_i \boldsymbol{g}_{ij} + [c, c].$$

Hence, maximizing $\sum_{i=1}^{m} (\boldsymbol{g}_{ij} + [c,c]) \succeq \boldsymbol{v}$ is equivalent to maximizing $\sum_{i=1}^{m} x_i \boldsymbol{g}_{ij} + [c,c] \succeq \boldsymbol{v}$. A similar result follows immediately for the column player.

Continuing, since the entries in $\boldsymbol{G}$ can be assumed to be positive, we have $\boldsymbol{v} > 0$. However, the width of $\boldsymbol{v}$, in general, can vary. To "normalize" the width of $\boldsymbol{v}$ in order to investigate a method for solving these interval systems, we will now assume that $\boldsymbol{v}$ is a degenerate interval; that is, the width of $\boldsymbol{v}$ is zero. Hence, $\boldsymbol{v}$ can be simultaneously viewed as an interval and real number. Thus, in this case, dividing each of the inequalities in constrained optimization problem (7.5) by $\boldsymbol{v}$ and treating the resulting quotients $x_k/\boldsymbol{v}$ as a new real-valued variable z_k, we notice that maximizing $\boldsymbol{v}$ is equivalent to minimizing

$$\frac{1}{\boldsymbol{v}} = \frac{\sum_{i=1}^{m} x_i}{\boldsymbol{v}} = \sum_{i=1}^{m} z_i,$$

since $\sum_{i=1}^{m} x_i = 1$. Therefore, constrained optimization problem (7.5) can be converted into an "interval" linear programming[4] problem:

$$\left\{\begin{array}{l} \text{Minimize } z_1 + z_2 + \cdots + z_m \text{ subject to} \\ \\ \quad z_1 \cdot \boldsymbol{g}_{11} + z_2 \cdot \boldsymbol{g}_{21} + \cdots + z_m \cdot \boldsymbol{g}_{m1} \succeq 1 \\ \quad z_1 \cdot \boldsymbol{g}_{12} + z_2 \cdot \boldsymbol{g}_{22} + \cdots + z_m \cdot \boldsymbol{g}_{m2} \succeq 1 \\ \qquad \qquad \vdots \\ \quad z_1 \cdot \boldsymbol{g}_{1n} + z_2 \cdot \boldsymbol{g}_{2n} + \cdots + z_m \cdot \boldsymbol{g}_{mn} \succeq 1 \\ \qquad \qquad z_1, z_2, \cdots, z_m \geq 0 \\ \\ \text{where the "1" is the interval } [1,1]\text{. After} \\ \text{this linear programming problem is solved} \\ \text{for the values } z_1, z_2, \ldots z_m\text{, the final val-} \\ \text{ues of } x_1, x_2, \ldots x_m \text{ and } \boldsymbol{v} \text{ can be quickly} \\ \text{found.} \end{array}\right\} \tag{7.7}$$

To optimize his strategy, the row player will attempt to find a strategy distribution $P^* = (p_1^*, p_2^*, \ldots, p_m^*)$ and a largest value for $\boldsymbol{v}$ so that, for any strategy distribution Q for the column player, we will have $P^* \boldsymbol{G} Q^T \succeq \boldsymbol{v}$ for a fixed relational membership value α, treating $\boldsymbol{v}$ as a trivial interval. In other words, the row player must solve this optimization problem (for a fixed relational membership value $0 < \alpha \leq 1$).

In a similar fashion, the column player will attempt to find a strategy distribution $Q^* = (q_1^*, q_2^*, \ldots, q_n^*)$ and a smallest value for $\boldsymbol{w} \geq 0$ so that, for

[4] This is not a linear optimization problem in the usual sense.

any strategy distribution P for the row player, we will have $P\boldsymbol{G}(Q^*)^T \preceq \boldsymbol{w}$ for the same membership value α. Therefore, the corresponding system will be

$$\left\{\begin{array}{r} \text{Maximize } z_1 + z_2 + \cdots + z_m \text{ subject to} \\ \\ z_1 \cdot \boldsymbol{g}_{11} + z_2 \cdot \boldsymbol{g}_{12} + \cdots + z_n \cdot \boldsymbol{g}_{1n} \preceq 1 \\ z_1 \cdot \boldsymbol{g}_{21} + z_2 \cdot \boldsymbol{g}_{22} + \cdots + z_n \cdot \boldsymbol{g}_{2n} \preceq 1 \\ \vdots \\ z_1 \cdot \boldsymbol{g}_{m1} + z_2 \cdot \boldsymbol{g}_{m2} + \cdots + z_n \cdot \boldsymbol{g}_{mn} \preceq 1 \\ z_1, z_2, \cdots, z_n \geq 0 \end{array}\right\} \tag{7.8}$$

The values of P^*, Q^*, v, and w are determined by solving these systems.

If each interval $\boldsymbol{g}_{ij}$ is interpreted as a trapezoidal fuzzy number, each of the two previous systems becomes a fuzzy linear programming problem with a crisp objective function and fuzzy constraints. Several techniques for solving such fuzzy systems have been developed, including [6]. These techniques define the notion of an (approximate) optimal solution in a fuzzy context. However, it is still worthwhile to develop direct techniques to solve interval linear programming problems, computing exact interval solutions whenever possible. Hence, we continue to address the development of such a general theory.

7.6 Solving Interval Inequalities

To solve the optimization problems described in the previous section, we determine general techniques for finding optima constrained by systems of interval inequalities.

7.6.1 Single Inequalities

We first consider the simplest case, namely to maximize the real value z subject to $z \cdot \boldsymbol{x} \preceq \boldsymbol{y}$, where each of $\boldsymbol{x}$ and $\boldsymbol{y}$ is a positive interval. Clearly, if both $\boldsymbol{x}$ and $\boldsymbol{y}$ are degenerate intervals, then the maximum value of z is $\boldsymbol{y}/\boldsymbol{x}$. Now, consider the case when at least one of $\boldsymbol{x}$ and $\boldsymbol{y}$ is not degenerate. Since we are using a fuzzy comparison operator for interval comparisons, we will consider the following restatement of this linear inequality problem:

$$\left\{\begin{array}{l} \text{Given } 0 < \alpha \leq 1 \text{ and intervals } \boldsymbol{x} \text{ and } \boldsymbol{y}\text{, find the maximum value} \\ \text{of } z \text{ where } z \cdot \boldsymbol{x} \preceq \boldsymbol{y} \text{ with membership value not less than } \alpha. \end{array}\right\} \tag{7.9}$$

We will represent the relationship between $z \cdot \boldsymbol{x}$ and $\boldsymbol{y}$ in a planar context, where an interval $\boldsymbol{v}$ is represented by the ordered pair $(m(\boldsymbol{v}), r(\boldsymbol{v}))$, where $m(\boldsymbol{v})$ is the midpoint of the interval and $r(\boldsymbol{v})$ is the radius of the interval.

Since this analysis considers only positive intervals (i.e., $m(\boldsymbol{v}) < r(\boldsymbol{v})$), the corresponding point in this coordinate system must lie below the diagonal in Figure 7.3.

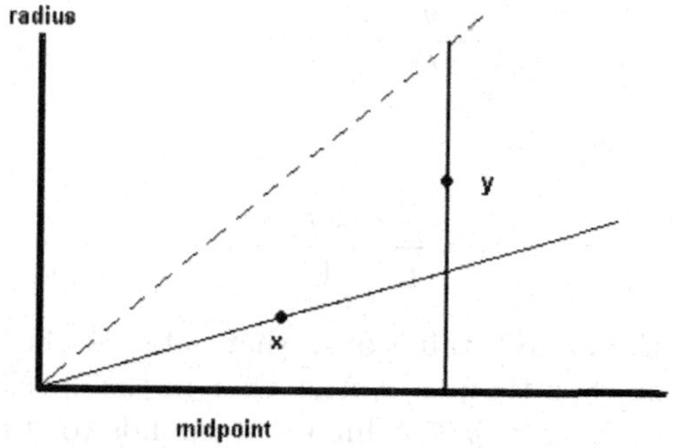

Fig. 7.3. Graphical Representation of $z \cdot \boldsymbol{x} \prec \boldsymbol{y}$.

Since the mapping $f(z) = z \cdot \boldsymbol{x}$ is linear,[5] it is easy to see that as z varies, the interval $z \cdot \boldsymbol{x}$ moves on the line from $(0,0)$ through $(m(\boldsymbol{x}), r(\boldsymbol{x}))$. The dynamics of how the interval $z \cdot \boldsymbol{x}$ "moves through" the interval $\boldsymbol{y}$ has three general cases that must be considered. To distinguish among these cases, consider the value of z for which the midpoint of $z \cdot \boldsymbol{x}$ equals the midpoint of $\boldsymbol{y}$. This value can easily be computed to be $(\underline{y} + \overline{y})/(\underline{x} + \overline{x})$, which we denote by c. One of three situations can occur for the relationship of $c \cdot \boldsymbol{x}$ to $\boldsymbol{y}$:

1. $c \cdot \boldsymbol{x} \subset \boldsymbol{y}$ and $c \cdot \boldsymbol{x} \neq \boldsymbol{y}$ (corresponds to the line from $(0,0)$ through $(m(\boldsymbol{x}), r(\boldsymbol{x}))$ in Figure 7.3 intersecting the vertical line containing $(m(\boldsymbol{y}), r(\boldsymbol{y}))$ below that point)
2. $c \cdot \boldsymbol{x} = \boldsymbol{y}$ (corresponds to the points $(0,0)$, $\boldsymbol{x}$ and $\boldsymbol{y}$ being collinear in Figure 7.3)
3. $\boldsymbol{y} \subset c \cdot \boldsymbol{x}$ and $c \cdot \boldsymbol{x} \neq \boldsymbol{y}$ (corresponds to the line from $(0,0)$ through $(m(\boldsymbol{x}), r(\boldsymbol{x}))$ in Figure 7.3 intersecting the vertical line containing $(m(\boldsymbol{y}), r(\boldsymbol{y}))$ above that point).

---

[5] It is worthy of note that $m(z\boldsymbol{x}) = zm(\boldsymbol{x})$ and $r(z\boldsymbol{x}) = zr(z)$ for real points z and intervals $\boldsymbol{x}$.

Consider the case $c \cdot \boldsymbol{x} = \boldsymbol{y}$. Clearly, $z = c$ is the maximum value as $c \cdot \boldsymbol{x} \leq \boldsymbol{y}$ crisply, and if $\epsilon > 0$, then $(c+\epsilon)\boldsymbol{x} \geq \boldsymbol{y}$ crisply so that $(c+\epsilon)\boldsymbol{x} \preceq \boldsymbol{y}$ has membership value 0.

Next, consider the case that $\boldsymbol{y} \subset c \cdot \boldsymbol{x}$ and $c \cdot \boldsymbol{x} \neq \boldsymbol{y}$. Hence, we see that $c\underline{x} < \underline{y}$ and $\overline{y} < c\overline{x}$. Since the membership values of $z \cdot \boldsymbol{x} \preceq \boldsymbol{y}$ is nonincreasing as z increases, we need only find the value of z such that the membership value of $z \cdot \boldsymbol{x} \preceq \boldsymbol{y}$ is equal to α. Hence, we to solve the equation

$$\frac{\underline{y} - z\underline{x}}{(\underline{y} - z\underline{x}) + (z\overline{x} - \overline{y})} = \alpha$$

for z. Doing so, one finds that

$$z = \frac{\underline{y} + \alpha(\overline{y} - \underline{y})}{\underline{x} + \alpha(\overline{x} - \underline{x})}.$$

Therefore, this is the largest value of z that satisfies the initial inequality with membership not less than α. Notice that in the special case of $\alpha = 1$, we get the optimal value $z = \overline{y}/\overline{x}$, which corresponds to the value where the left endpoints of $z \cdot \boldsymbol{x}$ and $\boldsymbol{y}$ are equal, which is where the crisp comparisons become fuzzy.

Considering the last case, namely $c \cdot \boldsymbol{x} \subset \boldsymbol{y}$ and $c \cdot \boldsymbol{x} \neq \boldsymbol{y}$; we once again must find the value of z so that the membership value of $z \cdot \boldsymbol{x} \preceq \boldsymbol{y}$. However, since $\boldsymbol{y}$ properly contains $z \cdot \boldsymbol{x}$ once the left endpoint of the two intervals agree, the portion of the interval $\boldsymbol{y}$ to the right of $z \cdot \boldsymbol{x}$ must be considered. Hence, in a symmetrical fashion to the previous case, the equation

$$\frac{\overline{y} - z\overline{x}}{(z\underline{x} - \underline{y}) + (\overline{y} - z\overline{x})} = \alpha$$

must be solved for z. Doing so generates the maximum value for z to be the expression

$$\frac{\overline{y} - \alpha(\overline{y} - \underline{y})}{\overline{x} - \alpha(\overline{x} - \underline{x})}.$$

Summarizing, we have the following theorem.

Theorem 3. *If each of $\boldsymbol{x}$ and $\boldsymbol{y}$ is a positive interval and $0 < \alpha \leq 1$, then there is a maximum value of the real-valued variable z such that $z \cdot \boldsymbol{x} \preceq \boldsymbol{y}$ with fuzzy membership value not less than α.*

Example 7. Solve the fuzzy linear programming problem for $\alpha = 0.9$:

$$\left\{ \begin{array}{c} \text{maximize } z \text{ subject to} \\ \\ z[1,2] \preceq [3,5] \\ z \geq 0 \end{array} \right\}$$

The value of z so that the midpoints are equal is $c = (3+5)/(1+2) = 8/3$. In this case, $[3,5]$ is a proper subset of $c[1,2] = [8/3, 16/3]$, so the maximum value of z that satisfies the inequality with the stated membership cut value is

$$z = \frac{3+0.9(2)}{1+0.9(1)} = \frac{4.8}{1.9} = 2.526315\ldots.$$

7.6.2 Extending to More General Cases

Let each of z_1 and z_2 be a real-valued variable, let each of $\boldsymbol{x}_1$, $\boldsymbol{x}_2$, and $\boldsymbol{y}$ be a positive interval, and fix α with $0 < \alpha \leq 1$. Consider the interval inequality

$$z_1 \cdot \boldsymbol{x}_1 + z_2 \cdot \boldsymbol{x}_2 \prec \boldsymbol{y}$$

and the objective function $z_1 + z_2$. Let the interval binary operator $\ominus$ be defined as $\boldsymbol{x} - \boldsymbol{y} = [\underline{x} - \underline{y}, \overline{x} - \overline{y}]$ provided $w(\boldsymbol{x}) \geq w(\boldsymbol{y})$. If z_1 is held constant between 0 and the corresponding maximum value of c that satisfies $c \cdot \boldsymbol{x}_1 \preceq \boldsymbol{y}$ (setting $z_2 = 0$ and solving the resulting simpler case using the fuzzy membership value α), then the maximum value of z_2 that satisfies the inequality $z_2 \cdot \boldsymbol{x}_2 \preceq (\boldsymbol{y} \ominus z_1 \cdot \boldsymbol{x}_1)$ using the membership value α can be determined by the above algorithm. The resulting value for z_2, in each of the three cases, is clearly a function of z_1; call it $z_{2_{\max(z_1)}}$. Hence, the original objective function can be rewritten as $z_1 + z_{2_{\max(z_1)}}$, which can be seen to be a continuous function of z_1. Therefore, the objective function must attain a maximum value on the interval $[0, c]$, which then can be used to determine the solution to the initial interval linear programming problem.

The following is a simple example that illustrates this approach.

Example 8. Solve the fuzzy linear programming problem for $\alpha = 0.9$:

$$\left\{\begin{array}{c} \text{maximize } x + y \text{ subject to} \\ \\ x[1,2] + y[2,3] \prec [4,8] \\ z \geq 0 \end{array}\right\}$$

Solution: We first consider the inequality $x[1,2] \prec [4,8]$. Note that the two intervals are collinear in the interval midpoint-radius plane, and setting the two midpoints equal gives $c = 4$. Therefore, we must consider the resulting inequality $y[2,3] \prec ([4,8] \ominus x[1,2])$ (i.e., $y[2,3] \prec [4-x, 8-2x]$), for each x in $[0,4]$. In the interval midpoint-radius plane, the interval $[2,3]$ lies below the line containing $[1,2]$ and $[4,8]$; hence, the line containing $(0,0)$ and the interval $[2,3]$ intersects the vertical line containing $[4-x, 8-2x]$ below that point. See Figure 7.4. Therefore, for each value of x in $[0,4]$, the corresponding value of y is

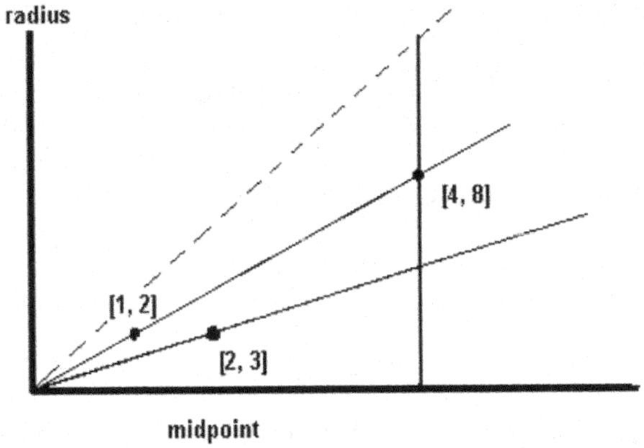

Fig. 7.4. $x[1,2] + y[2,3] \prec [4,8]$

$$y = \frac{(8-2x) - 0.9(8-2x-(4-x))}{3-0.9(3-2)} v = \frac{(2-0.9)(4-x)}{3-0.9} = \frac{1.1(4-x)}{2.1}.$$

We must optimize the objective function

$$x + y = x + \frac{1.1(4-x)}{2.1}$$

on the interval $[0,4]$. The derivative of this function is $1 - 1.1/2.1$, which is positive. Therefore, the maximum value of the objective function occurs when $x = 4$ and $y = 0$.

7.7 Conclusions

A model for crisply and fuzzily determined interval-valued Nash games has been developed using an appropriate fuzzy interval comparison operator. This model parallels the classical game context in a closely analogous way. Also, the theory of optimal mixed strategies for interval-valued games has been introduced, once again mimicking the classical model of converting the game into a linear programming problem.

To use interval linear programming techniques to find optimal mixed strategies in interval games, some assumptions must be made relative to the expected value interval $\boldsymbol{v}$. Assuming that this interval is degenerate generates corresponding "interval" linear programming problems that can be quickly solved. However, as the expected value of the game corresponds to a linear combination of the entries in the game matrix, this assumption appears to be unrealistic.

Acknowledgment: This work is partially supported by the U.S. National Science Foundation under grant CISE/CCF-0202042.

References

1. Allen, J.F.: Maintaining knowledge about temporal intervals. Communications of the ACM 26(11), 832–843 (1983)
2. Collins, D., Hu, C.: Studying interval valued matrix games with fuzzy logic. Soft Computing. 12(2), 147–155 (2008)
3. Dutta, P.K.: Strategies and Games: Theory and Practice. The MIT Press, Cambridge, MA (1999)
4. Garagic, D., Cruz, J.B.: An approach to fuzzy noncooperative Nash games. Journal of Optimization Theory and Applications 118(3), 475–491 (2003). URL `http://dx.doi.org/10.1023/B:JOTA.0000004867.66302.16`
5. Fishburn, P.C.: Interval Orders and Interval Graphs: A Study of Partially Ordered Sets. Wiley, New York (1985)
6. Fuller, R., Zimmermann, H.: Fuzzy reasoning for solving fuzzy mathematical programming problems. Fuzzy Sets and Systems 60, 121–133 (1993)
7. Nash, J.: Equilibrium points in n-person games. Proceedings of the National Academy of Sciences of the United States of America 36, 48–49 (1950)
8. Nash, J.: Non-cooperative games. The Annals of Mathematics 54(2), 286–295 (1951). URL `http://jmvidal.cse.sc.edu/library/nash51a.pdf`
9. Russell, S., Lodwick, W.A.: Fuzzy game theory and internet commerce: estrategy and metarationality. In: R.L. Muhanna, R.L. Mullen (eds.) Proceedings of the Annual Meeting of the North American Fuzzy Information Processing Society (NAFIPS), pp. 93–98. IEEE, New York (2002)
10. Winston, W.L.: Operations Research: Applications and Algorithms, 4th ed. Thomson Brooks/Cole, Pacific Grove, CA (2004)
11. Wu, S.H., Soo, V.W.: A fuzzy game theoretic approach to multi-agent coordination. In: PRIMA '98: Selected Papers from the First Pacific Rim International Workshop on Multi-Agents, Multiagent Platforms, pp. 76–87. Springer-Verlag, London (1999)
12. Zadeh, L.A.: Fuzzy sets. Information and Control 8(3), 338-353 (1965)

8

Interval-Weighted Graphs and Flow Networks

Chenyi Hu and Ping Hu

Computer Science Department, University of Central Arkansas, 201 Donaghey Avenue, Conway, AR 72035-0001, USA. chu,phu@uca.edu

Weighted graphs are useful computational models and have broad applications in decision making and knowledge processing. In contrast to the classical definition, weights in real-world applications often vary within intervals rather than being constant. In this chapter, we study interval-weighted graphs. By defining a fuzzy partial order relationship for intervals, we extend classical graph algorithms to interval-weighted graphs and capacity flow networks. An application on task management is modeled with an interval capacity flow network as an example.

8.1 Interval-Weighted Graphs

A graph $G = (V, E)$ consists of a set of vertices (V) and a set of edges (E). G is weighted if for every edge $e \in E$ there is a weight associated with e. In applications, these weights can represent meaningful things such as distance or cost. Therefore, weighted graphs have been well studied and broadly used in solving real-world applications.

A graph G is undirected if the two vertices of every edge are not ordered. Otherwise, it is a directed graph or digraph. A path of a graph is a consecutive sequence of edges. A graph G is connected if for any two vertices A and B in G there exists a path such that one can travel between A and B. The graphs studied in this chapter are initially assumed to be positively weighted, connected, and undirected.

In classical graph theory, the weights in a graph are constants. However, in real applications, these weights can vary within ranges rather than constants. For example, travel time from A to B may not be exactly 2 hours but usually between 1 hour and 50 minutes and 2 hours 5 minutes. As another example, the available bandwidth of a network connection may be between 75% and 80% during a given time period. To better model such variability of weights in a graph, instead of using constants, we represent weights as intervals. We

C. Hu et al. (eds.), *Knowledge Processing with Interval and Soft Computing*, DOI: 10.1007/978-1-84800-326-2_8,

investigate interval-weighted graphs in this chapter.[1] Our initial investigations into interval-weighted graphs are reported in [9]. Figure 8.1 presents an interval-weighted, connected, undirected graph.

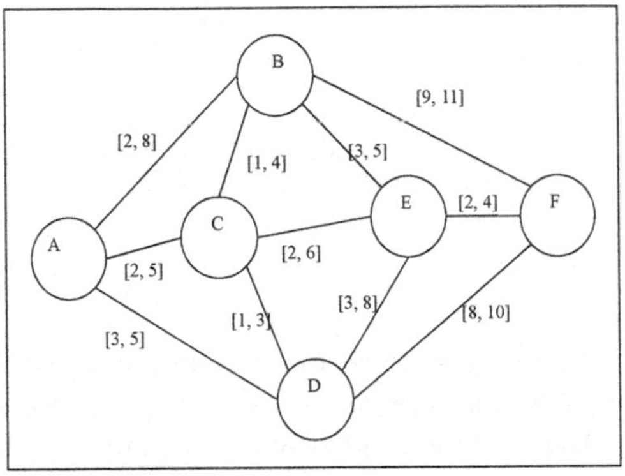

Fig. 8.1. An interval-weighted graph.

Typical applications of a weighted graph include finding shortest paths and identifying a minimum spanning tree. Algorithms for finding shortest paths include Dijkstra's algorithm [4], the Bellman-Ford algorithm [2], and others. Algorithms that find a minimum spanning tree include Kruskal's algorithm [11], the Prim-Jarnik algorithm [12] and Borůvka's algorithm [3].

In studying shortest paths and minimum spanning trees of weighted graphs, an ordering relationship is necessary to compare distances or the sum of the weights. To this end, we must investigate ways to order intervals for comparison.

8.2 Fuzzy Partial Order Relations for Intervals

8.2.1 Incomparability of Intervals in Binary Logic

For any two given real numbers x and y, based on their positions on the real line, the statement "x is less than y" can only be either true or false.

[1] The concept of interval-weighted graph is different from the term *interval graph* in existing literature. A definition of an interval graph can be found in [6], and stated as "It [an interval graph] refers a graph $(X, \sim)$ whose points can be mapped into intervals in a linearly ordered set such that, for all distinct x and y, $x \sim y$ if and only if the intervals assigned to x and y have a nonempty intersection."

However, the relation between two nonempty intervals $\boldsymbol{x}$ and $\boldsymbol{y}$ can be fairly complicated. They can be disconnected, partially overlapping, or completely overlapping. In [1], Allen listed 13 possible temporal relationships between 2 time intervals. Krokhin et al. further studied the relations in [10], and indicated that the relations between intervals could be $2^{13} = 8192$ possible unions of the 13 basic interval relations. This means that the statement "an interval $\boldsymbol{x}$ is less than another interval $\boldsymbol{y}$" cannot be expressed in binary logic. In short, there is not a general binary ordering relationship between two intervals.

8.2.2 A Binary Interval Operator

In studying interval matrix games in Section 7.2 of Chapter 7, we discussed ways to determine if an interval $\boldsymbol{x}$ is less than or equal to another interval $\boldsymbol{y}$ with the binary interval operator $\preceq$. We further modify the discussion and define a binary interval operator, $\prec$, to indicate the degree (or fuzzy membership) of an interval $\boldsymbol{x}$ less than another interval $\boldsymbol{y}$. Then we prove that the operator $\prec$ in fact establishes a fuzzy partial order relation for intervals.

Let $\boldsymbol{x}$ and $\boldsymbol{y}$ be two intervals. If $x < y, \forall x \in \boldsymbol{x}$ and $y \in \boldsymbol{y}$, we say that $\boldsymbol{x}$ is less than $\boldsymbol{y}$. This happens only when $\boldsymbol{x} \cap \boldsymbol{y} = \emptyset$ and $\boldsymbol{x}$ is completely on the left side of $\boldsymbol{y}$. In this case, we denote $\boldsymbol{x} \prec \boldsymbol{y} = 1$.

An interval $\boldsymbol{x}$ can be on the left of another interval $\boldsymbol{y}$ but partially overlapped (i.e., $\underline{x} \leq \underline{y} \leq \overline{x} < \overline{y}$). In this case, we may say that "$\boldsymbol{x}$ is weakly less than $\boldsymbol{y}$" and denote $\boldsymbol{x} \prec \boldsymbol{y} = 1^-$.

Assume $\boldsymbol{x} \subset \boldsymbol{y}$ and $\boldsymbol{x} \neq \boldsymbol{y}$; thus, $\underline{x} \leq \underline{y} \leq \overline{x} \leq \overline{y}$ but $\underline{x} = \underline{y}$ and $\overline{x} = \overline{y}$ cannot both be true simultaneously. It is easy to prove $0 \leq \dfrac{\overline{y} - \overline{x}}{w(\boldsymbol{y}) - w(\boldsymbol{x})} \leq 1$ when $\boldsymbol{x} \subset \boldsymbol{y}$ and $\boldsymbol{x} \neq \boldsymbol{y}$. Also, when $\underline{x} = \underline{y}$, $\dfrac{\overline{y} - \overline{x}}{w(\boldsymbol{y}) - w(\boldsymbol{x})} = 1$ (close enough to 1^-) and $\dfrac{\overline{y} - \overline{x}}{w(\boldsymbol{y}) - w(\boldsymbol{x})} = 0$ as $\overline{x} = \overline{y}$. Hence, we define

$$\boldsymbol{x} \prec \boldsymbol{y} = \frac{\overline{y} - \overline{x}}{w(\boldsymbol{y}) - w(\boldsymbol{x})}.$$

When the midpoints of $\boldsymbol{x}$ and $\boldsymbol{y}$ overlap (i. e., $m(\boldsymbol{x}) = m(\boldsymbol{y})$, and $w(\boldsymbol{x}) \neq w(\boldsymbol{y})$), we have $\dfrac{\overline{y} - \overline{x}}{w(\boldsymbol{y}) - w(\boldsymbol{x})} = 0.5$.

Finally, when $\boldsymbol{x}$ and $\boldsymbol{y}$ are the same, one is equally greater and less than the other, and we write $\boldsymbol{x} \prec \boldsymbol{y} = 0.5$.

Summarizing the above discussion, we define a binary operation with the operator $\prec$ for two intervals $\boldsymbol{x}$ and $\boldsymbol{y}$ as follows.

Definition 1. *Let $\boldsymbol{x} = (\underline{x}, \overline{x})$ and $\boldsymbol{y} = (\underline{y}, \overline{y})$ be two intervals and let $\prec$ be a binary interval operator. The binary operation $\boldsymbol{x} \prec \boldsymbol{y}$ returns a real between 0 and 1 as*

$$\boldsymbol{x} \prec \boldsymbol{y} = \begin{cases} 1 & \textit{if } \overline{x} < \underline{y} \\ 1^- & \textit{if } \underline{x} \le \underline{y} \le \overline{x} < \overline{y} \\ \dfrac{\overline{y} - \overline{x}}{w(\boldsymbol{y}) - w(\boldsymbol{x})} & \textit{if } \underline{y} \le \underline{x} < \overline{x} \le \overline{y} \textit{ and } w(\boldsymbol{x}) < w(\boldsymbol{y}) \\ 0.5 & \textit{if } w(\boldsymbol{x}) = w(\boldsymbol{y}) \textit{ and } \underline{x} = \underline{y} \end{cases} \tag{8.1}$$

Since the value of $\boldsymbol{x} \prec \boldsymbol{y}$ is between 0 and 1, it can be viewed as the fuzzy membership for the statement "$\boldsymbol{x}$ is less than $\boldsymbol{y}$". This definition also works when one or both of $\boldsymbol{x}$ and $\boldsymbol{y}$ are trivial intervals. The above definition implies the following corollaries.

Corollary 1. *Let $\boldsymbol{x}$ and $\boldsymbol{y}$ be two intervals, then the following holds:*

1. $\boldsymbol{x} \prec \boldsymbol{y} = 0.5$ *iff* $m(\boldsymbol{x}) = m(\boldsymbol{y})$.
2. $\boldsymbol{x} \prec \boldsymbol{y} > 0.5$ *iff* $m(\boldsymbol{x}) < m(\boldsymbol{y})$.
3. $\boldsymbol{x} \prec \boldsymbol{y} < 0.5$ *iff* $m(\boldsymbol{x}) > m(\boldsymbol{y})$.

Proof. We prove these three statements one by one.

1. Assume $\boldsymbol{x} \prec \boldsymbol{y} = 0.5$. If $\boldsymbol{x} = \boldsymbol{y}$, their midpoints are the same (i.e., $m(\boldsymbol{x}) = m(\boldsymbol{y})$). Otherwise, by Definition 1, $0.5 = \dfrac{\overline{y} - \overline{x}}{w(\boldsymbol{y}) - w(\boldsymbol{x})} = \dfrac{\overline{y} - \overline{x}}{2r(\boldsymbol{y}) - 2r(\boldsymbol{x})}$. Hence, $\overline{y} - \overline{x} = r(\boldsymbol{y}) - r(\boldsymbol{x}) = \dfrac{\overline{y} - \underline{y}}{2} - \dfrac{\overline{x} - \underline{x}}{2}$. Therefore, $\underline{y} + \overline{y} = \underline{x} + \overline{x}$ and $m(\boldsymbol{x}) = m(\boldsymbol{y})$.
   Now, assume $m(\boldsymbol{x}) = m(\boldsymbol{y})$. If $\boldsymbol{x} = \boldsymbol{y}$, from Definition 1, $\boldsymbol{x} \prec \boldsymbol{y} = 0.5$. If $\boldsymbol{x} \neq \boldsymbol{y}$, then $\underline{y} + \overline{y} = \underline{x} + \overline{x}$. Hence, $\overline{y} - \overline{x} = \underline{x} - \underline{y}$. However, $w(\boldsymbol{y}) - w(\boldsymbol{x}) = \overline{y} - \underline{y} - (\overline{x} - \underline{x}) = (\overline{y} - \overline{x}) + (\underline{x} - \underline{y}) = 2(\overline{y} - \overline{x})$. Hence, $\boldsymbol{x} \prec \boldsymbol{y} = \dfrac{\overline{y} - \overline{x}}{w(\boldsymbol{y}) - w(\boldsymbol{x})} = 0.5$.
2. Assume $\boldsymbol{x} \prec \boldsymbol{y} > 0.5$. If $\boldsymbol{x} \prec \boldsymbol{y} = 1$, then $\overline{x} < \underline{y}$. Since $m(\boldsymbol{x}) \le \overline{x}$ and $\underline{y} \le m(\boldsymbol{y})$, we have $m(\boldsymbol{x}) < m(\boldsymbol{y})$. If $\boldsymbol{x} \prec \boldsymbol{y} = 1^-$, then $\underline{x} \le \underline{y} \le \overline{x} < \overline{y}$. Hence, we have $\underline{x} + \overline{x} < \underline{y} + \overline{y}$. This implies $m(\boldsymbol{x}) < m(\boldsymbol{y})$. Otherwise, $\boldsymbol{x} \prec \boldsymbol{y} = \dfrac{\overline{y} - \overline{x}}{2r(\boldsymbol{y}) - 2r(\boldsymbol{x})} > 0.5$ implies $\overline{y} - \overline{x} > r(\boldsymbol{y}) - r(\boldsymbol{x})$ (i.e., $\overline{y} - \overline{x} > \dfrac{\overline{y} - \underline{y}}{2} - \dfrac{\overline{x} - \underline{x}}{2}$). Hence, $\underline{y} + \overline{y} > \underline{x} + \overline{x}$ and $m(\boldsymbol{x}) < m(\boldsymbol{y})$.
   Now assume $m(\boldsymbol{x}) < m(\boldsymbol{y})$. Then $\underline{x} + \overline{x} < \underline{y} + \overline{y}$ implies $\overline{y} - \overline{x} > r(\boldsymbol{y}) - r(\boldsymbol{x})$. If $\underline{y} \le \underline{x} < \overline{x} \le \overline{y}$ and $r(\boldsymbol{y}) > r(\boldsymbol{x})$, then $\boldsymbol{x} \prec \boldsymbol{y} = \dfrac{\overline{y} - \overline{x}}{2(r(\boldsymbol{y}) - r(\boldsymbol{x}))} > 0.5$. Otherwise, $\boldsymbol{x} \prec \boldsymbol{y} = 1$ or 1^-.
3. Assume $\boldsymbol{x} \prec \boldsymbol{y} < 0.5$. Then we have $\overline{y} - \overline{x} < r(\boldsymbol{y}) - r(\boldsymbol{x})$ (i.e., $\overline{y} - \overline{x} < \dfrac{\overline{y} - \underline{y}}{2} - \dfrac{\overline{x} - \underline{x}}{2}$). Hence, we have $\underline{x} + \overline{x} > \underline{y} + \overline{y}$. This implies $m(\boldsymbol{x}) > m(\boldsymbol{y})$.
   Now, assume $m(\boldsymbol{x}) > m(\boldsymbol{y})$. Then $\underline{x} + \overline{x} > \underline{y} + \overline{y}$ implies $\overline{y} - \overline{x} < r(\boldsymbol{y}) - r(\boldsymbol{x})$. Hence, $\boldsymbol{x} \prec \boldsymbol{y} = \dfrac{\overline{y} - \overline{x}}{2[r(\boldsymbol{y}) - r(\boldsymbol{x})]} < 0.5$.

Corollary 2. *Let* $\boldsymbol{x}$ *and* $\boldsymbol{y}$ *be two intervals and* $\boldsymbol{x} \prec \boldsymbol{y} \neq 1$. *Then* $\boldsymbol{x} \prec \boldsymbol{y} = (\boldsymbol{x}+\boldsymbol{z}) \prec (\boldsymbol{y}+\boldsymbol{z})$ *for a proper interval* $\boldsymbol{z}$.

Proof. From the definition, if $\boldsymbol{x} \prec \boldsymbol{y} = 1^-$, then $\underline{x} \leq \underline{y} \leq \overline{x} < \overline{y}$. Since $\underline{z} \leq \overline{z}$, we have $\underline{x} + \underline{z} \leq \underline{y} + \underline{z} \leq \overline{x} + \overline{z} < \overline{y} + \overline{z}$. Hence, $(\boldsymbol{x}+\boldsymbol{z}) \prec (\boldsymbol{y}+\boldsymbol{z}) = 1^-$.

If $\boldsymbol{x} \prec \boldsymbol{y} = 0.5$, then $m(\boldsymbol{x}) = m(\boldsymbol{y})$. Hence, $m(\boldsymbol{x}+\boldsymbol{z}) = m(\boldsymbol{x}) + m(\boldsymbol{z}) = m(\boldsymbol{y}) + m(\boldsymbol{z}) = m(\boldsymbol{y}+\boldsymbol{z})$, and $(\boldsymbol{x}+\boldsymbol{z}) \prec (\boldsymbol{y}+\boldsymbol{z}) = 0.5$.

Otherwise, $(\boldsymbol{x} \prec \boldsymbol{y}) = \dfrac{\overline{y} - \overline{x}}{w(\boldsymbol{y}) - w(\boldsymbol{x})}$. Since $(\overline{y} + \overline{z}) - (\overline{x} + \overline{z}) = \overline{y} - \overline{x}$ and $w(\boldsymbol{y}+\boldsymbol{z}) - w(\boldsymbol{x}+\boldsymbol{z}) = w(\boldsymbol{y}) - w(\boldsymbol{x})$, we have $(\boldsymbol{x} \prec \boldsymbol{y}) = (\boldsymbol{x}+\boldsymbol{z}) \prec (\boldsymbol{y}+\boldsymbol{z})$.

As a dual of the above discussion, we can define a binary operator $\succ$ as the following to indicate the degree of $\boldsymbol{x}$ greater than $\boldsymbol{y}$

Definition 2. *Let* $\boldsymbol{x}$ *and* $\boldsymbol{y}$ *be two intervals and let* $\succ$ *be a binary interval operator that returns the fuzzy membership of the statement "*$\boldsymbol{x}$ *is greater than* $\boldsymbol{y}$*" as* $(\boldsymbol{x} \succ \boldsymbol{y}) = 1 - (\boldsymbol{x} \prec \boldsymbol{y})$.

Similarly, we have the following corollary.

Corollary 3. *Let* $\boldsymbol{x}$ *and* $\boldsymbol{y}$ *be two intervals. Then*

1. $\boldsymbol{x} \succ \boldsymbol{y} = 0.5$ *iff* $m(a) = m(b)$.
2. $\boldsymbol{x} \succ \boldsymbol{y} > 0.5$ *iff* $m(a) > m(b)$.
3. $\boldsymbol{x} \succ \boldsymbol{y} < 0.5$ *iff* $m(a) < m(b)$.

8.2.3 Fuzzy Partial Order Relations for Intervals

In binary logic, a relation R on a set X is a partial order iff (a) $\forall x \in X, xRx \rightarrow$ false (inreflexive) and (b) $\forall x, y, z \in X, (xRy, yRz) \rightarrow xRz$ (transitive); then R is a partial order relation on X. To extend these concepts in fuzzy logic, we define the concepts of fuzzy inreflexibility, fuzzy transitivity, and fuzzy partial order relation as follows. We then prove that the binary operator $\prec$ is in fact a fuzzy partial order relation for intervals.

Definition 3. *A fuzzy relation* R *on a set* X *is fuzzily inreflexive if* $\forall x \in X, xRx = 0.5$; R *is fuzzily transitive if* $\forall x, y, z \in X$, *if* $xRy > 0.5$ *and* $yRz > 0.5$; *then* $xRz > 0.5$. *If* R *is both fuzzily inreflexive and transitive, then* R *is a fuzzy partial order relation.*

Theorem 1. *The binary interval operators* $\prec$ *is a fuzzy partial order relation for intervals.*

Proof. Since $\boldsymbol{x} \prec \boldsymbol{x} = 0.5$, the operator $\prec$ is fuzzily irreflexive.

Let $\boldsymbol{x}, \boldsymbol{y}$, and $\boldsymbol{z}$ be intervals. From Corollary 1, $\boldsymbol{x} \prec \boldsymbol{y} > 0.5$ implies $m(\boldsymbol{x}) < m(\boldsymbol{y})$, and $\boldsymbol{x} \prec \boldsymbol{y} > 0.5$ implies $m(\boldsymbol{y}) < m(\boldsymbol{z})$. The midpoints of intervals are just reals. Hence, $\boldsymbol{x} \prec \boldsymbol{y} > 0.5$ and $\boldsymbol{x} \prec \boldsymbol{y} > 0.5$ imply $m(\boldsymbol{x}) < m(\boldsymbol{z})$ and $\boldsymbol{x} \prec \boldsymbol{z} > 0.5$. Therefore, the binary operator $\prec$ is fuzzily transitive.

Hence, the binary interval operator $\prec$ is a fuzzy partial order relation for intervals.

Similarly, we can easily prove the interval operator $\succ$ forms a fuzzy partial order too.

We have now established fuzzy partial orders for intervals in terms of fuzzy membership. We complete this section with an example.

Example 1. For the two nested intervals $\boldsymbol{x} = [0, 4]$ and $\boldsymbol{y} = [1, 3]$, the fuzzy memberships for "$\boldsymbol{x}$ is less than $\boldsymbol{y}$" and "$\boldsymbol{x}$ is greater than $\boldsymbol{y}$" are both 0.5 since $m(\boldsymbol{x}) = m(\boldsymbol{y}) = 2$.

Letting $\boldsymbol{x} = [0, 3]$ and $\boldsymbol{y} = [0, 4]$, "$\boldsymbol{x}$ is less than $\boldsymbol{y}$" has a fuzzy membership of 1^-, whereas the fuzzy membership for "$\boldsymbol{x}$ is greater than $\boldsymbol{y}$" is zero.

For the intervals $\boldsymbol{x} = [1, 3]$ and $\boldsymbol{y} = [0, 5]$, '$\boldsymbol{x}$ is less than $\boldsymbol{y}$' has a fuzzy membership of 2/3, whereas the fuzzy membership of "$\boldsymbol{x}$ is greater than $\boldsymbol{y}$" is 1/3.

8.3 Shortest Paths and Minimum Spanning Trees for Interval-Weighted Graphs

Finding shortest paths and minimum spanning trees have be applied to knowledge processing and decision making. With the fuzzy partial order relations for intervals, we now study fuzzy shortest path and minimum spanning tree for interval-weighted graphs in this section.

8.3.1 Dijkstra's Shortest Path Algorithm

First, let us review the classical Dijkstra's shortest path algorithm [4] for a constant weighted graph. It is provided that all weights are positive. To find a shortest path from a vertex, v, to all others in a connected graph, Dijkstra's algorithm uses growing "clouds" with a greedy approach. Instead of repeating the details of the well-known Dijkstra's algorithm, we list it in pseudo-code here.

Algorithm 6 Dijkstra's Shortest Path Algorithm

```
Input: An undirected graph G with nonnegative weights, and
   a starting vertex v in G.
Output: A label D[u] for each vertex u of G, such that D[u]
   is the shortest distance from v to u in G.

for each vertex u in G
  if (u = v)
      D[u] = 0
  else
      D[u] = infinity

Store all vertices in a priority queue Q with D[u] as key
```

```
while  (Q is not empty)
  u = Pop(Q)
  for all vertexes z adjacent to u and z in Q do
    if D[u] + w(u, z) < D[z] then
       D[z] = D[u] + w(u, z)

Return the label D[u] of each vertex of G
```

The most critical steps of Dijkstra's algorithm are (1) to keep a priority queue for bringing in a vertex to the "cloud", and (2) to update the distance labels of `D[u] + w (u, z)` after a new vertex `u` is brought into the "cloud". Whenever `D[u] + w (u, z) < D[z]`, we update `D[z]` by `D[u] + w (u, z)` and then update the priority queue `Q`. This edge relaxation process guarantees that only the vertex with minimum distance is added in to the "cloud."

To extend classical Dijkstra's algorithm for interval-weighted graphs, we only need to apply the fuzzy partial order relation for intervals to maintain a priority queue and to perform the edge relaxation. For example, the condition for edge relaxation would be `if [(D[u] + w(u, z))` $\prec$ `D[z]]` > 0.5, `then D[z] = D[u] + w(u, z)`. Similarly, one may extend the Bellman-Ford algorithm [2] to find shortest paths of an interval-weighted digraph.

8.3.2 Crisp and Fuzzy Shortest Paths for an Interval-Weighted Graph

In an interval-weighted graph, the weight of a path P is defined as the total weight of its edges, that is,

$$w(P) = \sum_{i, e_i \in P} w(e_i).$$

A path P^* from a vertex A to another vertex B is *crisply* shortest if $w(P^*) \prec w(P) = 1$ or 1^- for all other paths P from A to B. Otherwise, if $0.5 < w(P^*) \prec w(P) < 1$ for all paths P from A to B, P^* is *fuzzily* shortest. In applying an interval shortest path for knowledge processing, one need to consider its fuzzy membership.

Applying Dijkstra's algorithm to an interval-weighted graph, we bring a vertex `u` into Dijkstra's "cloud" based on the least interval label `D[u]` that is the total interval weight from the starting vertex `v` to `u`. For example, if $\mathtt{u}_0$ is the first vertex brought into the "cloud," then the interval weight of the edge `v`-$\mathtt{u}_0$ is the least, either crisply or fuzzily, among all edges directly connected to the starting vertex `v`.

If every top element in the priority queue of Dijkstra's algorithm has crisply least interval label, clearly the shortest paths are crisp. Readers can easily verify that the shortest paths of the interval-weighted graph G in Figure 8.1 from A to C, E, F are crisp. The shortest path from A to C has an interval weight [2, 5]. The shortest path from A → C → E has an interval weight [4, 11]; and A → C → E → F has an interval weight [6, 15].

Importantly, when the label of the dequeued vertex `D[u]` is the least, even fuzzily, compared with labels of vertices other than `u`, we would prefer to bring in `u` into the "cloud" rather than others. Since all edges have positive weight, bringing in a vertex other than `u`, say `z`, would result in a longer path to `u`. This is because we must add at least one more edge from `z` to reach `u`.

However, there is a case that we should consider more carefully. There can be multiple paths to the top vertex `u` of the priority queue, and the label `D[u]` is fuzzily the least among those of these paths. In this case, bringing `u` into the cloud results in a fuzzy shortest path. The fuzzy membership of the label `D[u]` being the least represents the degree of the path is shortest. Furthermore, the fuzziness will be inherited to `z` when `z` is brought into the "cloud" through the edge (u, z). This is from Corollary 2.

Finally, the following theorem indicates the uniqueness of a shortest path if its fuzzy membership is greater than 0.5.

Theorem 2. *Let P be a path from a vertex A to another vertex B in an interval-weighted graph. If the fuzzy membership of P being the shortest is greater than 0.5, then there is no other path from A to B being the shortest with membership greater than 0.5.*

Proof. Let P and P' be two distinct paths from A to B and both have the fuzzy membership being the shortest greater than 0.5. Then $weight(P) \prec weight(P') > 0.5$ and $weight(P') \prec weight(P) > 0.5$. Since $\prec$ is a fuzzy partial order from Theorem 1, it is fuzzily transitive. Hence, $weight(P) \prec weight(P) > 0.5$. However, $\prec$ is also fuzzily irreflexive; that is, $weight(P) \prec weight(P) = 0.5$. Contradiction.

8.3.3 Minimum Spanning Tree for Interval-Weighted Graphs

By removing some edges from a connected graph $G = (V, E)$, one may form a subgraph $G' = (V, E')$ with the same number of vertices but less edges. If G' is in fact a tree T, then it is called a spanning tree of G. If G is weighted, then among all of its spanning trees the one with the minimum total weight is called the minimum spanning tree (MST) of G. Obviously, finding an MST can be useful in knowledge processing and for decision making.

The first algorithm that finds a minimum spanning tree was developed by the Czech scientist Borůvka [3]. Its purpose was to find efficient electrical coverage of Bohemia. Other algorithms to find the MST include Kruskal's MST algorithm [11] and Prim-Jarnik's MST algorithm [12]. Most MST algorithms, if not all, take a greedy approach. Hence, sorting is often required according to a partial ordering relation. The fuzzy partial order relation for intervals in this chapter can be applied to extend these MST algorithms for interval-weighted graphs.

To describe a practical algorithm that finds a minimum spanning tree for an interval-weighted graph, let us consider to extend the Kruskal's MST algorithm. For the reader's convenience, we list Kruskal's MST algorithm in pseudo-code.

Algorithm 7 Kruskal MST

```
Input: A connected weighted graph G
Output: A spanning tree T of G with the minimum total weight.

for each vertex v in G do
    define a Cloud(v) of {v}
let Q be a priority queue.
    Insert all edges into Q using their weights as the key
    T = empty
while T has fewer than n-1 edges do
    edge e = Pop(Q)
    Let u, v be the endpoints of e
      if Cloud(v) !=  Cloud(u) then
        Add edge e to T
        Merge Cloud(v) and Cloud(u)
return T
```

In the above algorithm, the most critical step is to construct a priority queue, Q, using the weights of edges as the key. Therefore, to extend Kruskal's MST algorithm to a connected interval weighted graph, we need to apply the fuzzy partial order relation for intervals to form a priority queue according to the interval weights of the edges.

If we use a heap [8] to implement the priority queue, the interval-weighted edges are stored in an AVL tree. Let M be the number of edges of the interval-weighted graph. The re-heap after a dequeue is at most $\log M$. Hence, the overall asymptotic complexity of the extended Kruskal's algorithm is still $\mathcal{O}(M \log M)$.

We now discuss the fuzzy membership of the MST generated by consecutively removing interval-weighted edges from the root of the heap. If a connected interval-weighted graph has N vertices, then its spanning tree consists of $N-1$ edges. The last dequeued edge from the root of the heap has its fuzzy membership being the least. By using e_{root}, e_{left}, and e_{right} to denote the edges in the root and its left and right children, respectively, the membership of e_{root} being the least is the smaller one between $e_{\text{root}} \prec e_{\text{left}}$ and $e_{\text{root}} \prec e_{\text{right}}$. It can be used as the fuzzy membership for the MST. If the last dequeued edge is the least interval of the heap crisply, then bringing any other edge in the rest of the heap would increase the total weight of the tree. Hence, the tree is an MST crisply. Otherwise, if the last dequeued edge is the least interval of the heap fuzzily, then the tree is an MST fuzzily. Similarly, we can extend other available MST algorithms for interval-weighted graphs with interval partial order relations.

Example 2. Find a minimum spanning tree in Figure 8.1.

In this simple example, instead of constructing an AVL tree to form a priority queue, we sort the edges according to their interval-weights in ascending order. They are: CD [1, 3], CB [1, 4], EF [2, 4], AC [2, 5], CE [2, 6], AD [3, 5], BE [3, 5], AB [2, 8], DE [3, 8], DF [8, 10], and BF [9, 11]. Applying the Kruskal's MST algorithm, we bring in CD, then CB, then EF, and then AC to construct the spanning tree without any questions. This results in two "clouds." One consists of A, B, C, and D. The other consists of E and F. To connect the two subtrees, we can pick BE [3, 5] or CE [2, 6] since $[3,5] \prec [2,6] = 0.5$. If we pick BE, the total weight of the spanning tree is [9, 21]. Otherwise, if we pick CE, the total weight is [8, 22]. Since $[9,21] \prec [8,22] = 0.5$, the two spanning trees are equally being minimum with membership 0.5.

8.4 Flow Networks with Interval Uncertainty

Capacity flow networks are useful in knowledge processing and decision making. We study them associated with interval uncertainty in this section. Before doing so, let us briefly review related concepts of a capacity flow network.

8.4.1 Capacity Flow Network

A directed graph $G = (V, E)$ can model a *flow network* with two specified nodes: S (source) and D (destination/sink). A flow on an edge $e \in E$ is denoted by $f(e)$. The maximum allowable flow on an edge $e \in E$ in the given direction is called the capacity of that edge and denoted as $c(e)$. A flow on the network is a function f that satisfies the capacity and conservation constraints described as $\forall e \in E : 0 \leq f(e) \leq c(e)$ and $\Sigma_{in} f(e) = \Sigma_{out} f(e), \forall v \in V$ except S and D. The value of a flow is the sum of all outgoing flow $f(e)$ from the source S.

Classically, the flow capacity on an edge, $c(e)$, is assumed a constant. Finding the maximum allowable flow, *max-flow* on a capacity network is a meaningful application. Ford and Fulkerson developed an algorithm [7] that could find a max-flow on a network computationally. The basic idea of the algorithm is as follows.

For a flow f on an edge $e = (u, v)$ directed from u to v, the forward residual capacity from u to v is denoted by $\Delta_f(u, v) = c(e) - f(e)$, where $c(e)$ is the forward capacity of e. The residual capacity from v to u, in the backward direction of the edge (u, v), is defined as $\Delta_f(v, u) = f(e)$.

Let π be a path from S to D that is allowed to traverse edges in either the forward or backward direction. The residual capacity $\Delta_f(\pi)$ of a path π is the minimum residual capacity of its edge; that is, $\Delta_f(\pi) = \min_{e \in \pi} \Delta_f(e)$. If $\Delta_f(\pi) > 0$, then π is called an *augment path.* A value of total flow then can be increased by adding the minimum residual capacity on each forward edge and subtracting it from every backward edge in the augment path.

By exhaustively finding augment paths on a capacity flow network, one may increase the total flow to the maximum within the capacity constraints. Figure 8.2 shows an example of a maximum flow on a capacity network. The flow $f(e)$ and capacity $c(e)$ associated with an edge e are represented compactly as $f(e)/c(e)$.

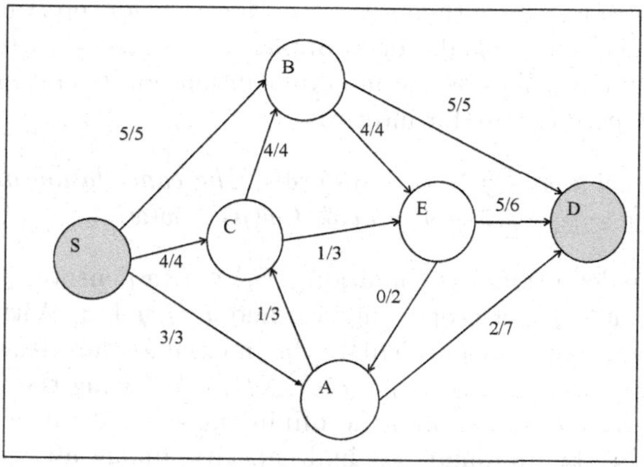

Fig. 8.2. Maximum flow on a capacity flow network

In [5], Edmonds and Karp then improved overall efficiency of the Ford-Fulkerson algorithm. By attaching a positive cost constant on each edge of a flow network, Edmonds and Karp further proposed their algorithm [5] that could find a max-flow with minimum cost on a capacity network.

Let C be a cycle, with both forward and backward edges, in a maximum flow network. For each $e \in C$, one can find its residual capacity. The product of the residual capacity and the cost on the edge is called the residual cost. The sign of a residual cost is determined as follows: positive if the edge is forward; otherwise negative. If the total residual cost on the cycle is negative, the cycle is called an augmented cycle. It has been proved that a max-flow is a min-cost flow if and only if it does not contain any augment cycle. Therefore, by repeatedly adjusting the flow on each edge of the augment cycle, the total cost will be reduced to approach the min-cost.

8.4.2 Considerations of Interval Uncertainty

In practice, due to uncertainty, the maximum capacity and the cost on an edge can be interval-valued. For example, a section of four-lane highway may open two to four lanes during a day because of construction. The cost of a flow on an edge may vary too. To model such kinds of uncertainty, we use intervals to represent flow capacity and cost on an edge. By allowing interval maximum capacity and cost, we have an *interval flow network.*

To shift a flow on an augment path and/or an augment cycle on a constant flow network, the real number arithmetic property of additive inverse is used. This means that if $b = c - a$, then $a + b = c$. However, for interval subtraction, this property is no longer valid. For example, $[1, 2] + [2, 4] = [3, 6]$, but $[3, 6] - [2, 4] = [-1, 4] \neq [1, 2]$. We use the interval cancellation operation, $\ominus$, defined below as an additive inverse for intervals.

Definition 4. *Let $\boldsymbol{x}$ and $\boldsymbol{y}$ be two intervals. The cancellation of $\boldsymbol{y}$ from $\boldsymbol{x}$ is defined as $\boldsymbol{x} \ominus \boldsymbol{y} = [\underline{x} - \underline{y}, \overline{x} - \overline{y}]$ provided $w(\boldsymbol{y}) \leq w(\boldsymbol{x})$.*

In the above definition, the condition $w(\boldsymbol{y}) \leq w(\boldsymbol{x})$ ensures $\underline{x} - \underline{y} \leq \overline{x} - \overline{y}$. Therefore, $\boldsymbol{z} = \boldsymbol{x} \ominus \boldsymbol{y}$ is a proper interval and $\boldsymbol{x} = \boldsymbol{y} + \boldsymbol{z}$. With the interval cancellation operation, we can calculate the interval forward residual capacity on the edge $e = (u, v)$ as $\Delta_f(u, v) = c(e) \ominus f(e)$. By using the fuzzy interval partial order relation, we can find the minimum residual capacity for a path π. Let $\Delta_f(m)$ be the minimum residual capacity among all edges in π with fuzzy membership greater than 0.5. Then it is the interval residual capacity of the path π. If $\Delta_f(m) \prec 0 > 0.5$, then π is an interval fuzzy augment path. We can then increase the total value of flow by adjusting flow on an interval fuzzy augment path. Similarly, we can find an interval fuzzy augment cycle to minimize the cost.

8.4.3 An Application: Job Scheduling on a Flow Network with Interval Uncertainty

As an application related to decision making, we study job/task scheduling on a capacity flow network. A job j is to start from the source S and to complete at the destination D. To simplify our study, we assume that the required capacity resource for completing a job is constant. However, the cost associated with each edge can vary within an interval. In practice, the cost may or may not be scalable. For example, the monetary charge of shipping a package from S to D can be proportional to its weight. However, driving a light sedan or an eighteen wheeler from S to D may take comparable time. In this application, we assume that the cost on each edge is proportional to the flux on that edge.

Max-Flow Min-Cost Scheduling

Let $J = \{j_1, j_2, \ldots, j_n\}$ be a collection of jobs. Assume that each job j_i requires a capacity f_i. We need to schedule them on a constant capacity network with an interval cost on each edge. Our objective is to fully utilize the capacity of the network and minimize the total cost. To meet the capacity constraint, we assume $\sum_{i=1}^{n} f_i \leq |f_{\max}|$, where $|f_{\max}|$ is the maximum capacity of the network.

The total cost of the flow f is $T(f) = \Sigma_{e \in E} t(e) f(e)$, where $t(e)$ denotes the unit cost of a flux on the edge e. To schedule J with minimum total interval cost, we first assign the jobs to the flow network and satisfy the capacity constraints. Then we try to shift flow from more cost paths to less until there is no such path available. Let C be a cycle, with both forward and backward edges, in a value $|f|$ flow network. For a forward edge $e \in C$, its residual cost interval, $R(e)$, is the product of its capacity residual and cost interval. If e is a backward edge, then its residual cost interval is the negative of the product. The residual cost interval of the cycle C, $R(C)$, is the sum of the residual cost interval of all edges on C, i. e., $R(C) = \Sigma R(e)$ for all $e \in C$.

Through interval computing, we can find the interval residual cost for the cycle C. By comparing $R(C)$ with zero, $R(C) \prec 0$, we obtain the degree of that the cycle C is a fuzzy augment cycle. If there is a flow that can be possibly shifted from a path p_1 to another path p_2 and reduces total cost, then p_1 and p_2 form an augment cycle with membership more than 0.5. By repeatedly adjusting flow on each edge of fuzzy augment cycle, the total interval cost will approach its minimum while maintaining the value of the flow. We summarize the above discussion as

Algorithm 8 Job scheduling on a capacity flow network with scalable interval cost

1. *Find the maximum capacity of the flow network* f_{max}.
2. *Assign the jobs to meet the capacity constraints, provided* $\Sigma f_i \leq f_{max}$.
3. *Repeatedly find augmenting cycles with fuzzy membership greater than 0.5 or a preset α-cut, and then reduce the total cost by shifting the flow.*

Job Scheduling with Cost Constraints

In the real-world, there can be a cost cap associated with a job to be assigned. To describe this, in addition to the capacity constraint, a job j can have a per unit cost constraint d_j associated with it. The overspending of completing j is the difference between its actual cost and d_j. Our objective of the scheduling is not only to minimize the overall cost but also minimize the sum of the possible overspendings.

To do this, we propose the following approach. We first find the max-flow minimum interval cost as described in Algorithm 8. Let $P = (p_1, p_2, p_3, \ldots, p_m)$ be paths from S to D of the flow with minimum total time costs and let $t(p_i)$

be the interval cost associated with the path p_i. We then sort these interval costs for all paths and obtain $t(p_1) \leq t(p_2) \leq \cdots \leq t(p_m)$ in terms of the fuzzy memberships. For n to-be-scheduled jobs, we can assume them in ascending order according to their cost caps. Now, we can assign the jobs according to their cost caps on sorted paths. We use a greedy approach to assign the job with the least cost cap to the available path with minimum costs. We illustrate the above ideas with the following example.

Example 3. Schedule three jobs j_1, j_2, and j_3 on the interval cost flow network illustrated in Figure 8.3. The capacity requirements and cost caps for the jobs are $f_1 = 5$, $d_1 = 21$; $f_2 = 2$, $d_2 = 14$; and $f_3 = 6$, $d_3 = 23$, respectively.

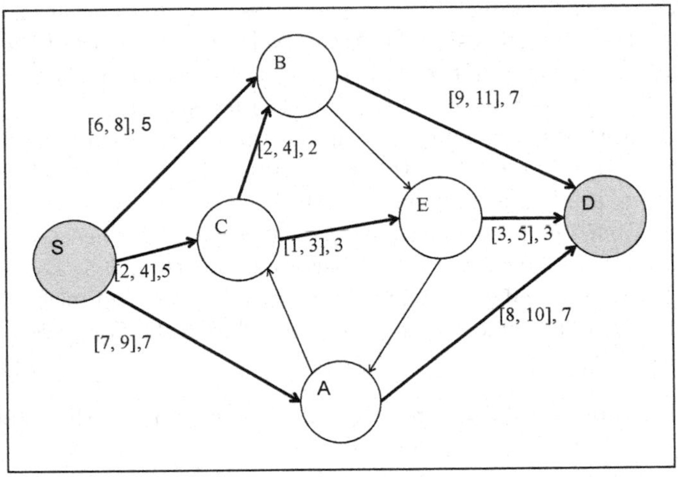

Fig. 8.3. Example: Task scheduling.

The required capacity to schedule the jobs is $f_1 + f_2 + f_3 = 5 + 2 + 6 = 13$. We find that the maximum allowable flow of the network is 17. Hence, the capacity constraint is satisfied. The paths and associated costs for the flow are as follows:

- $p_1 : S \to C \to E \to D$ with interval cost $t(p_1) = [2, 4] + [1, 3] + [3, 5] = [6, 12]$ and $f(p_1) = 3$.
- $p_2 : S \to C \to B \to D$ with interval cost $t(p_2) = [2, 4] + [2, 4] + [9, 11] = [13, 19]$ and $f(p_2) = 2$.
- $p_3 : S \to B \to D$ with interval cost $t(p_3) = [6, 8] + [9, 11] = [15, 19]$ and $f(p_3) = 5$.

- $p_4 : S \to A \to D$ with interval cost $t(p_4) = [7,9] + [8,10] = [15,19]$ and $f(p_4) = 7$.

We sort the paths according to their interval costs in the table below.

Path	Cost	Flux
p_1: S-C-E-D	[6, 12]	3
p_2: S-C-B-D	[13, 19]	2
p_3: S-B-D	[15,19]	5
p_4: S-A-D	[15,19]	7

We sort the jobs by their cost caps and get $d_2 < d_1 < d_3$. Thus, we first assign j_2 on p_1, which has the least cost. Now, p_1 is left with only one unit flow capacity available. We assign one unit of j_1 on p_1 and two units on each of p_2 and p_3. Finally, we assign j_3 on p_3 with three units and on p_4 with three units. Since p_3 and p_4 have the same interval cost, of course, we can assign all six units of j_3 on p_4. All tasks have been scheduled and meet the cost constraint.

There are cases in which some constraints cannot be crisply satisfied. For example, if one schedules j_i to a path p_j and $d_i \in t(p_j)$, then this assignment has a possibility to pass the cost cap due to interval cost uncertainty of the network. Applying the interval comparison operator that we defined earlier in this chapter, we can find the fuzzy membership. Furthermore, by calculate the center of gravity for such assignments, one can obtain a degree of belief for the possible total overcosts.

8.5 Conclusions

In this chapter, we introduced interval-weighted graphs and capacity flow networks that have broad applications in modeling real-world phenomena for knowledge processing and decision making. By extending constant weight, capacity, and cost to intervals, we are able to take interval uncertainty into our considerations and hence make reasonable decisions computationally.

The fuzzy partial order relations for intervals defined in this chapter have been the theoretical foundation for such extension. Using them, we can now compare intervals with a degree of belief. Applying these fuzzy partial order relations, we are able to find shortest paths and minimum spanning tree for interval-weighted graph. Furthermore, by using intervals to model uncertainties in capacity flow networks, we can find maximum flow and minimum cost for decision making. In addition to applications presented in this chapter, there should be more applications of the fuzzy partial order relations to other areas in knowledge processing.

Acknowledgment: This work is partially supported by the U.S. National Science Foundation under grants CISE/CCF-0202042 and CISE/CCF-0727798.

References

1. Allen, J.F.: Maintaining knowledge about temporal intervals. Communications of the ACM 26(11), 832–843 (1983)
2. Bellman, R.: On a routing problem. Quarterly of Applied Mathematics 16(1), 87–90 (1958). URL `http://wisl.ece.cornell.edu/ECE794/Jan29/bellman1958.pdf`
3. Borůvka, O.: On a certain minimal problem. Práce pravslé přírodovědecké společnosti 3, 37-58 (1926)
4. Dijkstra, E.W.: A note on two problems in connexion with graphs. Numerische Mathematik 1, 269–271 (1959). URL `http://jmvidal.cse.sc.edu/library/dijkstra59a.pdf`
5. Edmonds, J., Karp, R.: Theoretical improvements in the algorithmic efficiency for network flow problems. ACM 19, 248–264 (1972)
6. Fishburn, P.C.: Interval Orders and Interval Graphs: A Study of Partially Ordered Sets. Wiley, New York (1985)
7. Ford, L.R., Fulkerson, D.R.: Flows in Networks. Princeton University Press, Princeton, NJ (1962)
8. Goodrich, M., Tamassia, R.: Algorithm Design: Foundations, Analysis, and Internet Examples. Wiley, New York (2002)
9. Hu, P., Hu, C.: Fuzzy partial-order relations for intervals and interval weighted graphs. In: Proceedings of Foundations of Computational Intelligence, pp. 120-127 IEEE, New York (2007)
10. Krokhin, A., Jeavons, P., Jonsson, P.: Reasoning about temporal relations: The tractable subalgebras of Allen's interval algebra. Journal of the Association for Computing Machinery 50(5), 591–640 (2003)
11. Kruskal, J.B.: On the shortest spanning subtree of a graph and the traveling salesman problem. Proceedings of the American Mathematical Society 7(1), 48–50 (1956)
12. Prim, R.C.: Shortest connection networks and some generalizations. Bell System Technical Journal 36, 1389–1401 (1957)

9

Arithmetic on Bounded Families of Distributions: A DEnv Algorithm Tutorial

Daniel Berleant[1], Gary Anderson[2], and Chaim Goodman-Strauss[3]

[1] Department of Information Science, University of Arkansas at Little Rock, 2801 S. University Avenue, AR 72204, USA. jdberleant@ualr.edu
[2] Department of Applied Science, University of Arkansas at Little Rock, 2801 S. University Avenue, Little Rock, AR, 72204, USA. gtanderson@ualr.edu
[3] Department of Mathematics, University of Arkansas, Fayetteville, AR 72701, USA. strauss@uark.edu

Monte Carlo analysis is traditionally used in risk analysis to model uncertainty in the values of inputs of various kinds, such as initial conditions and variables. Although Monte Carlo has proven useful, extensive experience has revealed limitations in the technique. These limitations have motivated new techniques that overcome those limitations. This chapter focuses on an alternative approach: the DEnv algorithm. We begin by briefly discussing limitations of Monte Carlo simulation, followed by ways of attempting to address these limitations within the Monte Carlo paradigm. Then we discuss the DEnv (from **D**istribution **Env**elopes) algorithm, a technique for working with bounded families of probability distributions.

9.1 Motivation: Monte Carlo Simulation and Its Limits

It is useful to start with a critical look at Monte Carlo simulation, because the benefits of bounded families of distributions can best be appreciated in the context of the limitations of the traditional Monte Carlo method. There are two limitations that are especially significant in motivating use of bounded families of distributions in certain problems. These are described in the next two subsections.

9.1.1 When Knowledge Is Insufficient to Specify a Probability Distribution for a Model Variable

If some variable is uncertain, that uncertainty can often be modeled with a distribution function. However, if insufficient information exists to specify the exact shape of that distribution function, it is impossible to draw the samples needed by Monte Carlo simulation, unless either some distribution is

C. Hu et al. (eds.), *Knowledge Processing with Interval and Soft Computing*, DOI: 10.1007/978-1-84800-326-2_9,

arbitrarily assumed to apply or the variable is described by an interval instead of a distribution function. An arbitrary distribution (e.g., a normal or "bell" curve) might be assigned to a variable so that samples could be drawn from it, but although this would enable Monte Carlo analysis to proceed, the cost would be in making an unjustified assumption about the variable. Unjustified assumptions about model variables tend to imply results that, literally, are also unjustified.

To get justifiable results, given a variable with an unknown distribution, one might choose to bound it by using an interval to describe its range (possibly excluding the tails of the distribution if such a move is reasonable given the problem). Then this interval could be sampled, leading to results that bound the range of values of the outputs. However an interval is a relatively weak characterization of a variable that ignores information that may be available, such as variance and mean, that could potentially be used to help characterize the outputs.

Let us consider two examples. In the first, the information available is insufficient to specify a single distribution. In the second, an interval is suitable for describing uncertainty about the available information.

Example 1. Kolmogorov in [13] showed that a distribution obtained from a limited number of data points is likely to be significantly wrong and that confidence limits (in the form of bounds around the nominal distribution defined by the data) are more appropriate. Figure 9.1 shows an example of a distribution function and its confidence limits. Frame A, at the top, shows a cumulative distribution obtained from a set of data points. The s-curve shown rises unevenly due (one might reasonably speculate) to random noise in the data, although, in principle, the unevenness might actually accurately reflect the underlying random variable. Frame C, at the bottom, shows probability bounds that describe the confidence limits of the curve of frame A at the 0.9 probability level. The true distribution (which could be obtained from a limitless number of sample data) will fall within the bounds of the confidence limits with a probability of 0.9. Put another way, there is a 0.1 probability that the true distribution will cross outside the enveloping bounds shown at least once.

Example 2. A manufacturer of thermostats or some other measurement or control device might state limits on the measurement error and the controlled quantity. In this instance, information exists about the range of a variable but not about its distribution within that range. Intervals would be appropriate here for expressing uncertainty because they state lower and upper bounds. Continuing the thermostat case, if the temperature setting s is $67°$, the manufacturer states that settings are accurate to $\pm 1°$, and the manufacturer also specifies a hysteresis of $\pm 1°$ (i.e., the heater turns on at $s - 1 = 66°$ and turns off at $s + 1 = 68°$), then the actual temperature can be inferred to be within $(67 \pm 1 \pm 1)°$, or $[65, 69]°$. However its distribution within that range cannot be determined.

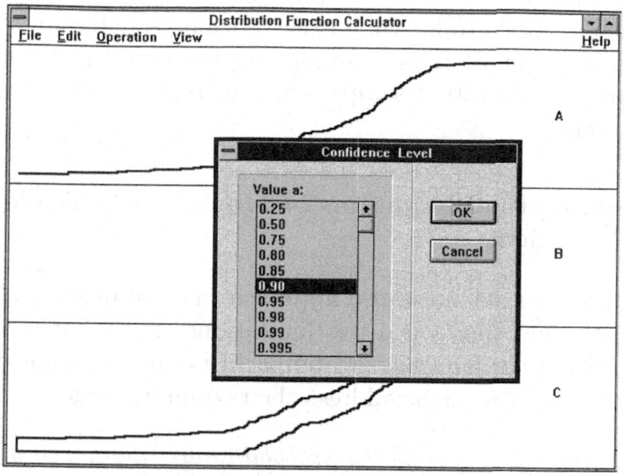

Fig. 9.1. A distribution derived from data, top, and its confidence limits, bottom (from [2].

9.1.2 Lack of Full Knowledge About the Dependency Relationships Among Variables

Suppose two variables have no significant relation to each other. For example, the price of oranges has no significant relation to the number of sunny days per year in Seattle, WA, USA. If distribution functions are available for both variables, each may be sampled to provide pairs of numbers to use in a simulation model, without fear that the value of a sample of one variable affects the distribution function that should be used to generate samples of the second variable. This is convenient both in implementability and in ease of modeling.

Another convenient situation is if the value of one variable completely determines the value of a second variable. For example, the number of sunny days per year completely determines the number of nonsunny days per year, and both values might be used in a simulation model of, say, utilization of tourist attractions (of which some would be more attractive on sunny days and others on nonsunny days). In this situation a sample drawn from one variable determines the sample to use of the other, and a simulation requiring both variables is both relatively simple and implementable.

However, a third situation often occurs that presents a problem. For example, an agricultural model of production might incorporate as variables both the price of the product and the number of sunny days in growing areas. It is likely that the values of those variables will be related in some way (i.e., not independent) but that this dependency is less than total (i.e., the value of one does not fully determine the other). Unless the joint probability distribution is known, which amounts to knowing exactly what distribution to sample for one variable given what value was sampled for the other, it is not possible in general to properly generate a sample value of one variable given a sample value of the other.

9.1.3 Overcoming the Limitations of Monte Carlo While Staying Within the Paradigm

Let us look at how the Monte Carlo approach may be made usable in situations in which the just-mentioned two limitations occur. Later this will help illustrate the advantages of a better approach, bounded families of distributions. The next two subsections address the two limitations.

Knowledge insufficient to specify the probability distribution of a model variable

Often a generic "reasonable" distribution will be used to model such a variable (e.g., a normal distribution). This permits a Monte Carlo model to be fully specified and therefore a simulation to be run. However, such an unjustified assumption about a model variable of course decreases the dependability of conclusions drawn from simulating the model.

Because potentially untrue assumptions can lead to problematic conclusions, we might wish to express only actual facts about variables. For example, we might model variables as intervals (i.e., ranges extending from minimum to maximum plausible values). If we expressed all variables this way, then a Monte Carlo simulation could be performed based on picking sample points randomly within those intervals for each simulation run. The results from many simulation runs would then be combined to give intervals describing ranges for the outputs. Unfortunately, it would say nothing about the shapes of their distributions, merely giving estimates of the ranges of their supports.

What if only some variables needed to be described using intervals because distributions were available for the others? Simulating Monte Carlo models that mix some variables that are interval-valued and others that are distribution-function-valued is less straightforward than if all were intervals or all were distributions. It would be easier to substitute, for each distribution in such a mixed model, an interval bounding the range of values permitted by the distribution. A disadvantage of this approach is that using intervals for variables for which distribution functions are known means ignoring available information. Although the conclusions drawn may be sufficient in some situations, they will tend to be weaker than if the distribution information available was used instead of ignored.

Example 3. We model whether a robotic vehicle can pull a cart containing cargo up a slope without its wheels slipping against the slope surface (Fig.

9.2), rendering it unable to complete its task [1]. This example can be applied to specific situations such as a robot moving cargo from an airplane drop to a central location, cargo transportation in rough and/or dangerous terrain, autonomous construction of bridges or other structures, and so on.

Fig. 9.2. Pioneer AT robot pulling a loaded cart up a hill.

The frictional force between the surface and the drive wheels of the robot must exceed the gravitational force pulling the cart down the incline (see Fig. 9.2). The force of gravity on the cart is $m_{cart} * g * \sin\theta$. Let us assume that the weight of the cargo-carrying cart is much higher than that of the robot. Then the force of friction on the wheels of the robot is $\mu_{friction} * m_{robot} * g * \cos\theta$. So, for the robot to successfully pull the cart up the incline requires that

$$\mu_{friction} * m_{robot} * g * \cos\theta > m_{cart} * g * \sin\theta.$$

In other words,

$$m_{robot} > \frac{(m_{cart} * \tan\theta)}{\mu_{friction}}$$

must hold. Given m_{robot}, the unknowns are $\mu_{friction}, m_{cart}$, and θ. These might be roughly estimated visually and from experience by the robot: m_{cart} from the size of the cargo; $\mu_{friction}$ by the color and glossiness of the incline's surface; and θ from stereo vision estimates of its depth at the top and bottom. These estimates, however, will have large uncertainties associated with them. Consider just one of these unknowns, m_{cart}, and define $m_{cart} = m_{vehicle} +$

m_{cargo}. Based on the manufacturer's specifications and known variabilities (e.g., how worn the tires are), $m_{vehicle}$ is likely to be known within an interval, such as $k \pm 10\%$ for some k. Given a sufficiently sized database of actual robot missions, m_{cargo} could be represented by a probability distribution. Therefore, to calculate m_{cart}, one must add two quantities, one of which is an interval and the other a distribution. An easy way to do this is to substitute a second interval for the distribution, thereby making the sum easy to calculate but losing potentially valuable information. Another easy way out would be to substitute a conjectured distribution for the interval, such as a triangular distribution with mode k. However, this has another problem, which is that the results of a simulation based on this model would be undependable because it uses a distribution that is merely a conjecture. The ideal approach would be to add an interval and a distribution. The problem is that a Monte Carlo model is not well suited to such a situation. In general it would require combining multiple Monte Carlo simulations, one for each of a random sample of values from the interval. (Well-behaved models could be handled by sampling only the endpoints of the interval.)

Lack of Knowledge About the Precise Nature of the Dependency Relationships Among Model Variables

The Monte Carlo approach may proceed straightforwardly if variables are assumed independent. However, if variables are not known for sure to actually be independent, resulting conclusions can be suspect. This is illustrated by the following example.

Example 4. Consider the case of a model with two uncertain variables that must be combined. Some bat populations have suffered fluctuations in population in recent years due to such factors as pesticides in their diet of insects, other human disturbance of habitats, and perhaps other poorly understood factors. In order to estimate the number of bats of a particular species in a particular area that will be present 1 year from now, one can add to the current population the product of the current population and the growth rate (a negative growth rate would signify a decline in population). Neither the current population nor the growth rate is likely to be known accurately and, therefore, might be better modeled using distributions than point values. Thus, we would have to multiply two distributions together to get an estimate of next year's population. Furthermore, the dependency relationship between these two distributions is unknown. They could be completely independent (which could lead to eventual extinction for negative growth rates). Alternatively, they could be positively correlated. This can occur in populations that are sufficiently low to be marginally viable. In that case, an increase in population can cause an increase in growth rate. Finally, it is possible for population and growth rate to be negatively correlated. This can occur, for example, when a population approaches the limit of the ability of the environment to support it, at which point individuals are forced into competition with each other for food and perhaps other resources, making it harder for them to survive and reproduce.

Thus, a model might specify distribution functions for population and growth rate, but the dependency relationship between the two distribution functions may be unknown. What, then, can be said about the product of population and growth rate and hence about the population in a year? If we assumed the distributions were independent, then the result of multiplying them would be some distribution. On the other hand, if we assumed the distributions were completely correlated (so a higher value for one implied a correspondingly higher value for the other), or negatively correlated (a higher value for one implied a correspondingly lower value for the other), then the result in each case would again be some distinct distribution.

Since we cannot justify any particular dependency relationship in this example, the result could be any of a family of distributions, each one corresponding to some dependency relationship - whether simple or complex - between the variables. Then the family of all possible result distributions, which includes independence, full positive and negative correlation, and all other dependency relationships, may be expressed using a bounded family of distributions to represent the space within which each member of that infinitely numerous family must be [3].

The sensitivity of any conclusions to an independence assumption can be checked, to a degree, by also running a Monte Carlo simulation on the problem under the assumption that the variables are perfectly positively correlated, as well as under the assumption that some are perfectly negatively correlated with others. These different assumptions, representing extremes of possible dependencies, will lead to possibly differing conclusions (although not necessarily to extremes within the space of conclusions implied by the space of possible dependencies; see Ferson et al. [9]). This will help test the sensitivity of the conclusions to assumptions about dependency.

The trustworthiness of a Monte Carlo simulation will generally be benefited when the dependencies among the variables are known. Correlations might be known even when full details of dependencies are not. If a correlation between two variables is positive, then a relatively high sample value for one variable would typically increase the probability of a relatively high sample value of the other variable. Similarly, a negative correlation would typically increase the probability of drawing a relatively low value of the other variable. The term "typically" applies because a positive correlation can hide a tendency for some high values of one variable to occur with low values of the other, if that tendency is overcompensated by a tendency for other high values of one to occur with high values of the other [9]. Even when correlation is known and modeled, underlying details about a dependency relationship that are hidden by the crude measure of correlation could impact the validity of the model and hence dependability of the results of a Monte Carlo simulation.

9.2 How Bounded Families of Distributions Can Help

We have just described how Monte Carlo simulation can be facilitated through unsupported assumptions (modeling an interval as a distribution or assuming a dependency relationship), or discarding information (as when modeling a distribution with an interval), or kludgey second-order modifications of the clean classical Monte Carlo approach. Ideally though, elegant techniques would be used that do not lead to reductions in information quality [17]). The approach described next, the DEnv technique, meets that requirement.

We begin by reviewing salient features of probability distributions. Because of their familiarity, they form a convenient lead-in to a discussion of bounded families of distributions.

The probability is 0 that a sample drawn from a probability density function will be less than the lowest value in its support, and it is 1 that it will be no greater than the greatest value in its support (Fig. 9.3). More generally, the probability that a sample will not exceed a specific value increases as the value specified increases. Based on that observation, a curve that plots probability against progressively increasing given values is called a cumulative distribution function (CDF), often abbreviated simply as "distribution" (Fig. 9.3b).

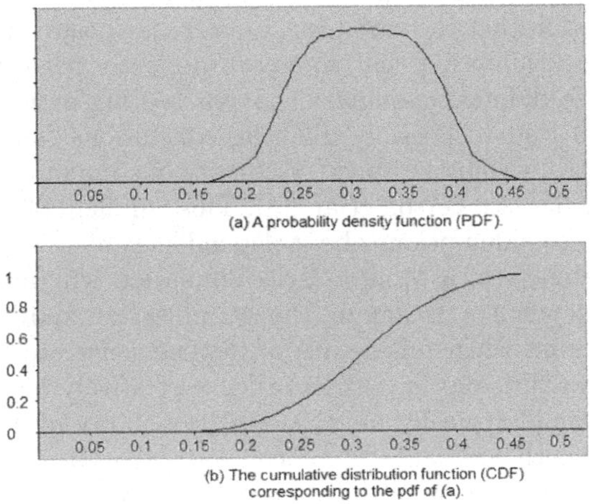

Fig. 9.3. A probability density function (PDF) and its corresponding cumulative distribution function (CDF). A CDF describes the cumulative area under its corresponding PDF, rising to a final value of 1. The capacity of these isomorphic representations to describe uncertainty is limited, motivating more general methods.

What happens if something can be said about a density function but not enough to specify it fully? For example, the mean and variance might be known, but not the detailed form of the curve. In such a situation, a family of different density functions conforms to the limited information we have about it. Almost any density function can be shifted right or left until its mean is a given value and then stretched or compressed around the mean to adjust its variance to another given value. Such a family of curves, if many were superposed, would form a jumble and be difficult to work with. Fortunately, this apparent jumble can be expressed in the more visualizable way discussed next.

9.3 Bounded Families of Distributions

If we integrate each member of a family of density functions to get a corresponding family of distributions, it is considerably easier to visualize and work with. Figure 9.4 shows envelopes bounding a family of distributions. This family corresponds to the family of all density functions with a given mean and variance. The envelopes shown, one bounding the family on the left and one on the right, are the bounds on this family of distributions.

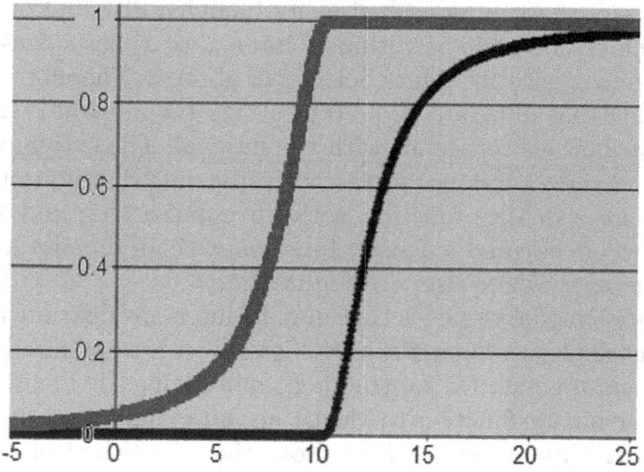

Fig. 9.4. Bounds around the family of cumulative distributions with mean 10 and variance 5. All such CDFs fall within these bounds, and some CDF in the family touches any given point on each bound. However, the bounding envelope curves do not themselves have mean 10 and variance 5. (The tails taper off to $\pm\infty$, not shown.)

Clearly, distribution family envelopes provide easily visualized bounds on the space through which members of the family can travel. Therefore, they also implicitly bound the corresponding family of density functions, which, as noted, is not as easy to visualize directly.

Let us next show that bounded distribution families enable a general strategy for circumventing the problems of traditional Monte Carlo simulation described earlier. What is needed is a representation for uncertainty that can (1) *express* intervals, distributions, and families of distributions and (2) *manipulate* model variables thus represented.

1. *Expressing* intervals, distributions, and bounded families of distributions. All of these can be expressed using bounded distribution families as a unifying representation, as we see next.
   a) *Families of distributions* are described using bounds as explained earlier. In principle, there are families that cannot be described with bounds (e.g., the family of all density functions with a single impulse and zero density everywhere else). In practice, bounding envelopes can represent the kinds of family of distributions that seem to typically arise in practice.
   b) *Distributions* are described using bounding envelopes easily, because a distribution is simply a family of distributions with one member. The appropriate bounds consist of a left envelope and a right envelope that are identical and equal to the distribution in question.
   c) *Intervals* are easily described using bounded distribution families. A variable restricted to be within an interval $[\underline{x}, \overline{x}]$ has a density function with zero density for values below $\underline{x}$ or above $\overline{x}$. Therefore, any density function that integrates from 0 to 1 over the interval is in the family of distributions consistent with the interval. The extremes giving the left and right envelopes of the corresponding distribution family are therefore a density function with an impulse at $\underline{x}$ and zero density everywhere else and a density function with an impulse at $\overline{x}$ and zero density everywhere else. See Figure 9.5.
2. *Manipulating* model variables that may be intervals, distributions, or families of distributions. Once the variables we wish to manipulate (e.g., by adding them together, or subtracting, multiplying, dividing, or applying some other binary function to them) are all expressed as bounded families of distributions, we need no more than a method of manipulating these bounded families. In other words, the conceptual differences among intervals, distributions, and distribution families become irrelevant to the manipulation method. We address such a method next.

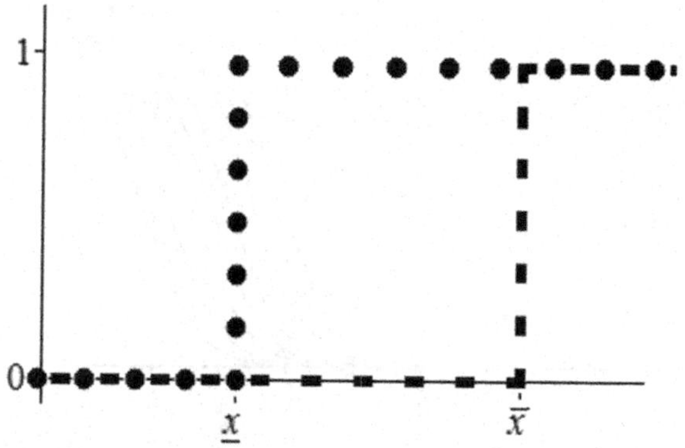

Fig. 9.5. Bounded family of distributions whose left and right envelopes (shown with dots and dashes, respectively) represent the interval $[\underline{x}, \overline{x}]$ (with low bound $\underline{x}$ and high bound $\overline{x}$).

9.4 Arithmetic Operations on Bounded Families of Distributions

We begin by showing how to apply a binary operation (e.g., addition) to variables when one is a distribution and the other is an interval. We will then extend the ideas to the other three cases of interest: one variable an interval and the other a bounded family of distributions, one a distribution and the other a bounded family, and both bounded families.

Consider the case where one variable is described using a distribution function and another is less well characterized, being described only by an interval describing its range of plausible values. The presence of a variable x described by an interval typically prevents representing the sum, product, and so on of x and some distribution, as a distribution. As Figure 9.6 shows, each possible value of the interval, when combined with the distribution leads to a distinct distribution for the output variable. Each distinct distribution is the distribution of the sum, given some particular sample value from the interval. The result is a family of distributions, one for each value in the interval. This family may be bounded with envelopes.

Up to this point, bounded distribution families have been illustrated - literally - graphically. But computer software for working with these families needs to represent them using numbers instead. The next section presents a method for calculating with bounded distribution families that can be implemented in computer software.

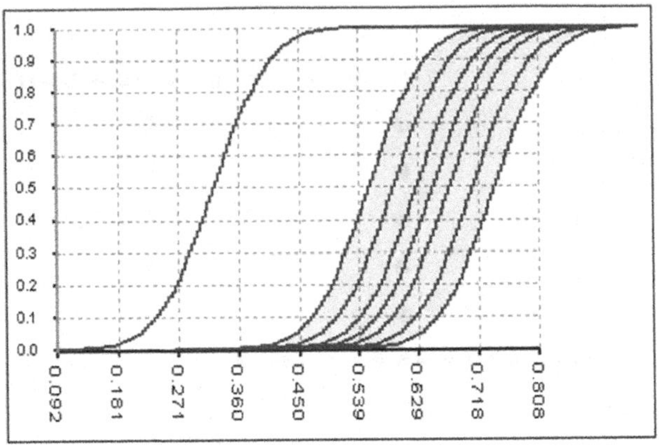

Fig. 9.6. The distribution farthest to the left is added to the interval [0.228, 0.421]. The result is the bounded family of distributions on the right, of which the left and right envelopes and five example interior members are shown. In this situation none of members of the family cross each other (but in other situations, they do).

9.5 A Numerical Approach to Computing with Bounded Distribution Families

A suitable way to work with bounded families of distributions on computers uses sets of *intervals* associated with *probabilities*. An interval, for this purpose, is a range described by its low and high endpoints. The interval containing all numbers from 2 through 9.8, for example, is written [2, 9.8]. In this approach, each interval is associated with the probability that a sample value of a random variable will belong to that interval. Graphically, a rectangle can be placed on the x-axis with its left and right sides at the low and high bounds of the interval, which has area = probability, so height = area/width. Figure 9.7 shows an example of a set of intervals with associated probabilities, its rectangles, and its bounded distribution family.

The rectangles are misleading in an important respect: They suggest that the distribution of probability for a given interval is uniform, because the interval and its probability are depicted using a rectangle with a flat top. In fact, no constraint on how probability is distributed within an interval is intended. At one extreme, probabilities might be concentrated as impulses at the low bounds of their intervals (i.e., at the left sides of the rectangles). Then the distribution family envelope curve will rise suddenly at the low bound of each interval (see the left staircase curve of Fig. 9.7), yielding the left envelope of a bounded distribution family, which is the fastest-rising curve

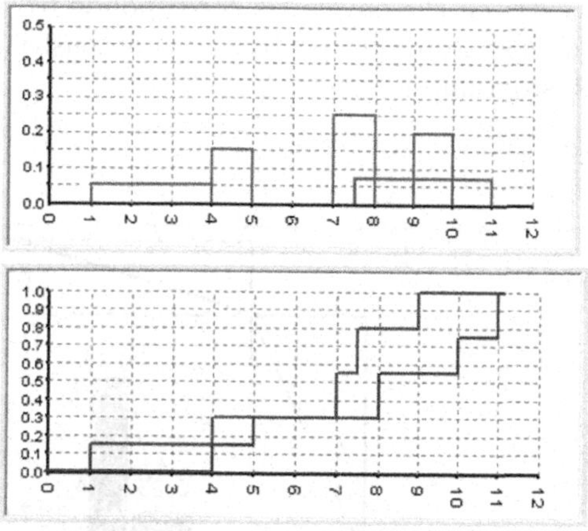

A partially specified distribution

```
{p([1,4])    = 0.15
 p([4,5])    = 0.15
 p([7,8])    = 0.25
 p([7.5,11])= 0.25
 p([9,10])   = 0.2
}
```

Fig. 9.7. An underspecified distribution consisting of a set of intervals and their associated probabilities, the corresponding rectangles, and the bounded distribution family.

that is consistent with the set of intervals and their probabilities. The opposite extreme would be to concentrate the probabilities at the high bounds of their corresponding intervals. Then the cumulative curve will rise suddenly at the high bounds of the intervals yielding a staircase curve that is the right-hand envelope of the bounded distribution family.

The left and right staircase-shaped envelope curves are bounding in that all curves that result from distributions of probabilities within their associated intervals travel between the two envelopes, never crossing them.

For example, Figure 9.8 shows envelopes and two other distributions that are consistent with those envelopes. One distribution has three straight segments, each corresponding to a uniform distribution within one of the histogram bars in the inset. Thus, when the probability of each interval is distributed uniformly over the interval (as suggested by the flat tops of the histogram bars in the inset), the cumulation rises in a series of connected, non-vertical line segments between the envelopes (dark middle curve). The smooth s-curve also shown between the envelopes corresponds to some smooth density function that the histogram discretizes. Such a density function is shown superposed on the histogram.

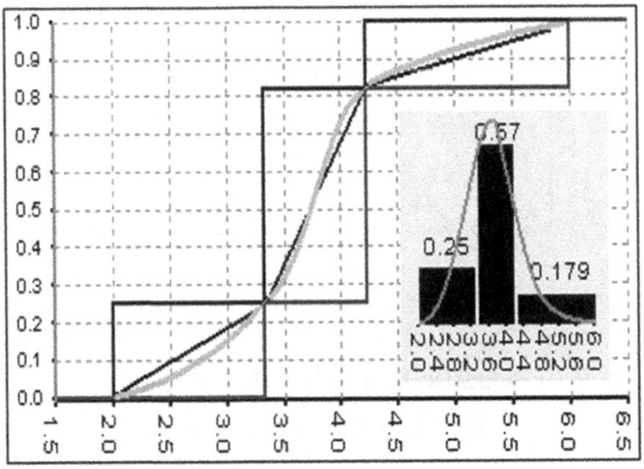

Fig. 9.8. Left and right staircase-shaped envelopes. In general, these envelopes may touch at one or more points, but they never cross. Within those bounds, two CDFs are shown: one composed of three straight line segments and one an s-curve. The inset shows rectangles arranged in a histogram and, superposed, an example of a density function that the histogram discretizes.

Showing distribution family bounds avoids problems with collections of rectangles, such as flat tops, which are misleading in seeming to suggest that probabilities are distributed uniformly over their intervals. Another potentially misleading visual characteristic of rectangle collections is that rectangles that overlap may lead to ambiguity regarding the identities of the intervals underlying them.

Although showing rectangles has limitations, so does showing bounding envelopes. Different sets of intervals and their probabilities can yield the same envelopes. For example, consider the sets S_1 and S_2:

$$S_1 = \{p([1,4]) = 0.5, p([2,3]) = 0.5\},$$
$$S_2 = \{p([1,3]) = 0.5, p([2,4]) = 0.5\}.$$

In both of those sets, the extreme case of concentrating probabilities at the low bounds of their intervals yields two impulses: one at 1 and the other at 2. The other extreme case, concentrating the probabilities at the high bounds of the intervals, also yields two impulses, one at 3 and the other at 4. Thus, the envelopes for S_1 and S_2 are identical. Yet, a random variable governed by S_1 can lead to different results than one governed by S_2. See Section 3.1 of [4] for a fuller discussion.

9.6 Discretization and Bounded Families of Distributions

Envelopes can be used for representing intervals, distributions, and bounded families of distributions. Also, we have been building the case that a suitable underlying data structure for expressing a pair of envelopes is a set of intervals and their associated probabilities. To further make this case requires addressing how to express smoothly curving distributions and envelopes using these sets. Intervals have definite endpoints and, graphically, sets of them yield left and right, sharply angled staircase like envelope curves. These are decidedly not smoothly curving. Yet, sets of intervals and probabilities can be used for representing smoothly curving envelopes as well. Figure 9.8 inset, shows a coarse discretization (three rectangles forming a histogram) of a curved density function. Histograms will approximate a probability density curve better as the number of histogram bars increases and their widths decrease. In cumulative terms, a distribution will be better approximated by its enclosing staircase shaped envelopes the more steps the envelopes possess. See Figure 9.9.

At this point we have introduced families of distributions with three alternative representations: (1) sets of intervals and probabilities, (2) rectangles, and (3) envelopes. The sets of intervals and probabilities are the underlying, computer-friendly specification, whereas rectangles and envelopes are human-friendly and derivable from the sets. We have not yet shown, however, how to take two different variables, each a bounded family of distributions, and add, subtract, multiply, or divide them or perform some other binary operation on them. This is discussed in the next section.

9.7 Computing with Bounded Distribution Families

We introduce how to do arithmetic computations on bounded distribution families using an example with a typical structure but artificial data.

Example 5. Consider the goal of finding out what can be determined about the total amount of some pesticide released into the environment worldwide. Model this total as C, where $C = A + B$, A is the amount contributed by U.S. agriculture, and B is the amount contributed by all other countries. The exact values of A and B are unknown, but we assume that distributions for them are available. We can discretize such distributions visually as histograms or as left and right envelopes (similar to those in Fig. 9.9), or alternatively for

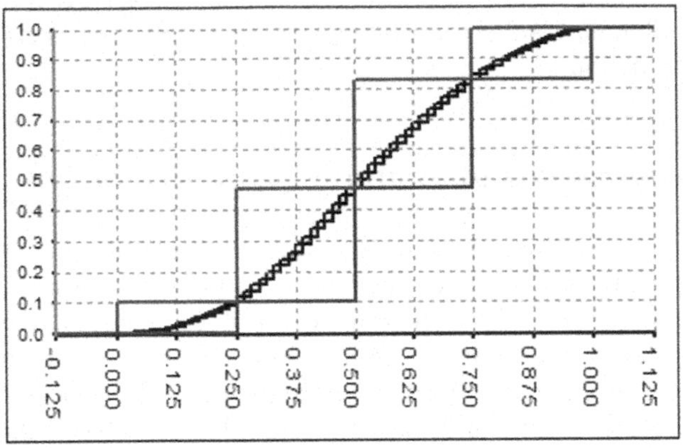

Fig. 9.9. Two discretizations of the same distribution, a light-colored pair of envelopes with 4 steps each, and a dark pair with 64. Each bar of a histogram that discretizes a density function corresponds in the world of distributions (the integrals of density functions) to a box, of which the north and east sides are formed by the left envelope and the south and west sides are formed by the right envelope.

computational purposes as sets of intervals and their probabilities. Table 9.1 shows the description in terms of intervals and probabilities.

A	B
p(A in [10,11])=0.1	p(B in [5,6])=0.05
p(A in [11,12])=0.2	p(B in [6,7])=0.06
p(A in [12,13])=0.4	p(B in [7,8])=0.08
p(A in [13,14])=0.2	p(B in [8,9])=0.1
p(A in [14,15])=0.1	p(B in [9,10])=0.21
	p(B in [10,11])=0.21
	p(B in [11,12])=0.1
	p(B in [12,13])=0.08
	p(B in [13,14])=0.06
	p(B in [14,15])=0.05

Table 9.1. The distribution functions describing the amounts contributed by pesticide sources A and B have been discretized and are shown symbolically as sets of intervals with associated probabilities.

To compute $A+B$ to get the total amount of pesticide released, consider that for any plausible value of A, B might potentially have any value permitted by its distribution. In terms of the intervals of Table 9.1, we add each interval in A to each interval in B to get $5 * 10 = 50$ new intervals, and we calculate a probability for each of the new intervals, resulting in a set of intervals and their probabilities for $C = A + B$. Thus, if A is in $[10, 11]$ and B is in $[5, 6]$, then C would be in $[10, 11] + [5, 6] = [15, 17]$. Similarly, we can get an interval describing the value of C given A in any of its 5 intervals and B in any of its 10 intervals. See Table 9.2.

A $B \rightarrow$ $\downarrow$	$[5, 6]$ $p = .05$	$[6, 7]$ $p = .06$	$[7, 8]$ $p = .08$	$[8, 9]$ $p = .1$	$[9, 10]$ $p = .21$	$[10, 11]$ $p = .21$	$[11, 12]$ $p = .1$	$[12, 13]$ $p = .08$	$[13, 14]$ $p = .06$	$[14, 15]$ $p = .05$
$[10, 11]$ $p = .1$	$[15, 17]$	$[16, 18]$	$[17, 19]$	$[18, 20]$	$[19, 21]$	$[20, 22]$	$[21, 23]$	$[22, 24]$	$[23, 25]$	$[24, 26]$
$[11, 12]$ $p = .2$	$[16, 18]$	$[17, 19]$	$[18, 20]$	$[19, 21]$	$[20, 22]$	$[21, 23]$	$[22, 24]$	$[23, 25]$	$[24, 26]$	$[25, 27]$
$[12, 13]$ $p = .4$	$[17, 19]$	$[18, 20]$	$[19, 21]$	$[20, 22]$	$[21, 23]$	$[22, 24]$	$[23, 25]$	$[24, 26]$	$[25, 27]$	$[26, 28]$
$[13, 14]$ $p = .2$	$[18, 20]$	$[19, 21]$	$[20, 22]$	$[21, 23]$	$[22, 24]$	$[23, 25]$	$[24, 26]$	$[25, 27]$	$[26, 28]$	$[27, 29]$
$[14, 15]$ $p = .1$	$[19, 21]$	$[20, 22]$	$[21, 23]$	$[22, 24]$	$[23, 25]$	$[24, 26]$	$[25, 27]$	$[26, 28]$	$[27, 29]$	$[28, 30]$

Table 9.2. Intervals for A are shown down the left, for B across the top, and for $C = A + B$ in the interior cells. For example, when $A \in [10, 11]$ and $B \in [5, 6]$, then $A + B \in [15, 17]$, and so on.

What about the probabilities associated with the interior cells of the table? These are not shown in the table because they vary for different dependency relationships between A and B.

We will call a table like Table 9.2 a *joint distribution tableau*. It shows the ranges of intervals for $C = A + B$ for all of the combinations of intervals in A and B. The probabilities associated with those intervals for C are constrained by the marginal probabilities, as shown in the first row and column, but are not fully determined because there is no information available about the dependency relationship between A and B. If A and B were independent, the probabilities of the interior cells would be the product of their marginal probabilities. However, A and B might not be independent. For example, heavy use of the pesticide in the United States might positively correlate with heavy use elsewhere due to similar judgments of farmers worldwide. On the other hand, if overall supply was limited, then heavy use in one country would limit its use elsewhere, a negative correlation. Each dependency relationship results in some distribution for C. We wish to construct the left and right envelopes around the family of all distributions plausible for C. Let us consider next a few selected values of the left and right envelopes bounding $C = A+B$, starting with the left envelope.

9.7.1 Left Envelope

$C = 14$: There is no way for $A + B$ to be as low as 14 no matter which of the possible values of A and B occur. Indeed, the lowest possible value of $A + B$ is 15, which would only occur if A and B were both at their minimum possible values of 10 and 5, respectively. Thus the left envelope height is zero at $C = 14$ (and all other values of C below 15). For the same reason, this is also true of the right envelope.

$C = 15$: The only way C can be 15 is if $A = 10$ and $B = 5$, which occurs only for the top left cell in the interior of the table. The probabilities in all of the interior cells in the row containing that cell sum to 0.1, because that is the marginal probability for $A \in [10, 11]$. Similarly, the probabilities associated with all of the interior cells in the *column* holding this cell must sum to 0.05 because that is the marginal probability for $B \in [5, 6]$. This puts an upper bound on the probability associated with the top left interior cell, of $\min(0.05, 0.1) = 0.05$. Some dependency relationship between A and B might be associated with such an assignment of probability to that cell but no dependency relationship can exist for which that probability would exceed 0.05. One might guess that this upper bound of 0.05 is not achievable because of a putative need to reserve some of the 0.05 marginal probability to distribute among other cells in that column. However, simply filling in probability values in the table by hand and adjusting them by trial and error reveals that in this case, the full 0.05 probability can be allocated to the top left interior cell (Table 9.3). Later we will discuss allocating probabilities automatically.

A $B \rightarrow$ $\downarrow$	$[5, 6]$ $p = .05$	$[6, 7]$ $p = .06$	$[7, 8]$ $p = .08$	$[8, 9]$ $p = .1$	$[9, 10]$ $p = .21$	$[10, 11]$ $p = .21$	$[11, 12]$ $p = .1$	$[12, 13]$ $p = .08$	$[13, 14]$ $p = .06$	$[14, 15]$ $p = .05$
$[10, 11]$ $p = .1$	$[15, 17]$ $p = .05$	$[16, 18]$ $p = .05$	$[17, 19]$ $p = 0$	$[18, 20]$ $p = 0$	$[19, 21]$ $p = 0$	$[20, 22]$ $p = 0$	$[21, 23]$ $p = 0$	$[22, 24]$ $p = 0$	$[23, 25]$ $p = 0$	$[24, 26]$ $p = 0$
$[11, 12]$ $p = .2$	$[16, 18]$ $p = 0$	$[17, 19]$ $p = .01$	$[18, 20]$ $p = 0.08$	$[19, 21]$ $p = .1$	$[20, 22]$ $p = .01$	$[21, 23]$ $p = 0$	$[22, 24]$ $p = 0$	$[23, 25]$ $p = 0$	$[24, 26]$ $p = 0$	$[25, 27]$ $p = 0$
$[12, 13]$ $p = .4$	$[17, 19]$ $p = 0$	$[18, 20]$ $p = 0$	$[19, 21]$ $p = 0$	$[20, 22]$ $p = 0$	$[21, 23]$ $p = .2$	$[22, 24]$ $p = 0.2$	$[23, 25]$ $p = 0$	$[24, 26]$ $p = 0$	$[25, 27]$ $p = 0$	$[26, 28]$ $p = 0$
$[13, 14]$ $p = .2$	$[18, 20]$ $p = 0$	$[19, 21]$ $p = 0$	$[20, 22]$ $p = 0$	$[21, 23]$ $p = 0$	$[22, 24]$ $p = 0$	$[23, 25]$ $p = 0$	$[24, 26]$ $p = .1$	$[25, 27]$ $p = .08$	$[26, 28]$ $p = .02$	$[27, 29]$ $p = 0$
$[14, 15]$ $p = .1$	$[19, 21]$ $p = 0$	$[20, 22]$ $p = 0$	$[21, 23]$ $p = 0$	$[22, 24]$ $p = 0$	$[23, 25]$ $p = 0$	$[24, 26]$ $p = 0.01$	$[25, 27]$ $p = 0$	$[26, 28]$ $p = 0$	$[27, 29]$ $p = .04$	$[28, 30]$ $p = .05$

Table 9.3. Joint distribution tableau for $C = A+B$. Each interior cell shows a range and probability for C associated with an interval and probability for B at the head of its column and for A at the head of its row. Many other probability asignments are also possible, corresponding to different dependencies between A and B.

Thus, the left envelope jumps from 0 to 0.05 at $C = 15$. This value cannot rise further until $C = 16$ because at that value for C, other interior cells associated with other ranges of A and B can contribute their probabilities to the ways in which C can be 16, as described next.

$C = 16$: The three cells whose summed probability we need to maximize in this case are the top left interior cell (call it the corner cell for now), the cell to its right, and the cell below it. As in the previous case, the leftmost interior column probabilities must add to $p(B = [5, 6]) = 0.05$. Also, the second column of interior cells must add up to 0.06. Both of those marginal probabilities can be distributed among just those three cells, leaving other interior cells in the two leftmost columns with zero probabilities. This maximizes the summed probability of those three cells at 0.11. Manual inspection of the problem is one way to reveal that such an allocation is possible (Table 9.4). This maximum probability of 0.11 applies for values of C from 16 up to 17, at which point other interior cells can contribute their probabilities to the ways in which C can be 17.

A ↓ \ B →	[5, 6] $p = .05$	[6, 7] $p = .06$	[7, 8] $p = .08$	[8, 9] $p = .1$	[9, 10] $p = .21$	[10, 11] $p = .21$	[11, 12] $p = .1$	[12, 13] $p = .08$	[13, 14] $p = .06$	[14, 15] $p = .05$
[10, 11] $p = .1$	[15, 17] $p = .04$	[16, 18] $p = .06$	[17, 19] $p = 0$	[18, 20] $p = 0$	[19, 21] $p = 0$	[20, 22] $p = 0$	[21, 23] $p = 0$	[22, 24] $p = 0$	[23, 25] $p = 0$	[24, 26] $p = 0$
[11, 12] $p = .2$	[16, 18] $p = .01$	[17, 19] $p = 0$	[18, 20] $p = .08$	[19, 21] $p = .1$	[20, 22] $p = .01$	[21, 23] $p = 0$	[22, 24] $p = 0$	[23, 25] $p = 0$	[24, 26] $p = 0$	[25, 27] $p = 0$
[12, 13] $p = .4$	[17, 19] $p = 0$	[18, 20] $p = 0$	[19, 21] $p = 0$	[20, 22] $p = 0$	[21, 23] $p = 0$	[22, 24] $p = .21$	[23, 25] $p = .1$	[24, 26] $p = .08$	[25, 27] $p = .01$	[26, 28] $p = 0$
[13, 14] $p = .2$	[18, 20] $p = 0$	[19, 21] $p = 0$	[20, 22] $p = 0$	[21, 23] $p = 0$	[22, 24] $p = .2$	[23, 25] $p = 0$	[24, 26] $p = 0$	[25, 27] $p = 0$	[26, 28] $p = 0$	[27, 29] $p = 0$
[14, 15] $p = .1$	[19, 21] $p = 0$	[20, 22] $p = 0$	[21, 23] $p = 0$	[22, 24] $p = 0$	[23, 25] $p = 0$	[24, 26] $p = 0$	[25, 27] $p = 0$	[26, 28] $p = 0$	[27, 29] $p = .05$	[28, 30] $p = .05$

Table 9.4. Three interior cells contribute to the cumulative probability $p(C \le 16)$. These are clustered in the upper left.

$C = 17$: To get the value of the left envelope at this value, we need to maximize the sum of the probabilities in six cells (clustered in the top left area of Table 9.5) whose intervals contain any values equal to 17 or less. Table 9.5 manages to allocate probabilities so that all of the probabilities in the first three interior columns are allocated within those six cells. The sum of the probabilities associated with those cells, 0.19, is maximal because any more probability would violate the constraints imposed by the marginal values of probability for B shown along the top of the table. This maximized probability applies over $17 \le C < 18$.

We can continue to work out the values of the left envelope for higher and higher values of C, but a naïve, pencil-and-paper approach gets unwieldy for mid-range values of C. Furthermore, computers do not use pencil and paper but require a well-defined procedure. Before discussing such a procedure, however, let us get a start on the right envelope, illustrating its nature as the dual of the left.

9.7.2 Right Envelope

$C = 21 - \epsilon$: To get the cumulative probability defining a y-axis value of a point on the right bounding envelope of C, we must *minimize* the sum of the probabilities associated with interior cells whose interval high bounds are

A ↓ $B \rightarrow$	[5, 6] $p = .05$	[6, 7] $p = .06$	[7, 8] $p = .08$	[8, 9] $p = .1$	[9, 10] $p = .21$	[10, 11] $p = .21$	[11, 12] $p = .1$	[12, 13] $p = .08$	[13, 14] $p = .06$	[14, 15] $p = .05$
[10, 11] $p = .1$	[15, 17] $p = \mathbf{0}$	[16, 18] $p = \mathbf{0}$	[17, 19] $p = \mathbf{.08}$	[18, 20] $p = .02$	[19, 21] $p = 0$	[20, 22] $p = 0$	[21, 23] $p = 0$	[22, 24] $p = 0$	[23, 25] $p = 0$	[24, 26] $p = 0$
[11, 12] $p = .2$	[16, 18] $p = \mathbf{0}$	[17, 19] $p = \mathbf{.06}$	[18, 20] $p = 0$	[19, 21] $p = .08$	[20, 22] $p = .06$	[21, 23] $p = 0$	[22, 24] $p = 0$	[23, 25] $p = 0$	[24, 26] $p = 0$	[25, 27] $p = 0$
[12, 13] $p = .4$	[17, 19] $p = \mathbf{.05}$	[18, 20] $p = 0$	[19, 21] $p = 0$	[20, 22] $p = 0$	[21, 23] $p = .15$	[22, 24] $p = .2$	[23, 25] $p = 0$	[24, 26] $p = 0$	[25, 27] $p = 0$	[26, 28] $p = 0$
[13, 14] $p = .2$	[18, 20] $p = 0$	[19, 21] $p = 0$	[20, 22] $p = 0$	[21, 23] $p = 0$	[22, 24] $p = 0$	[23, 25] $p = .01$	[24, 26] $p = .1$	[25, 27] $p = .08$	[26, 28] $p = .01$	[27, 29] $p = 0$
[14, 15] $p = .1$	[19, 21] $p = 0$	[20, 22] $p = 0$	[21, 23] $p = 0$	[22, 24] $p = 0$	[23, 25] $p = 0$	[24, 26] $p = 0$	[25, 27] $p = 0$	[26, 28] $p = 0$	[27, 29] $p = .05$	[28, 30] $p = .05$

Table 9.5. The cumulative probability for $C \leq 17$ is maximized by assigning probabilities to interior cells as shown. The probabilities contributing to the sum are bold. The full marginal probabilities of those columns may be assigned to the three interior cells holding the interval [17, 19], so their summed probability of $0.05 + 0.06 + 0.08 = 0.19$ is the maximum possible cumulation at $C = 17$.

below C and that therefore *must* contribute all of their probability to the cumulation. Every other interior cell holds an interval with a high bound above C, and so either *cannot* contribute probability to C (if its low bound is also above C) or *might not* contribute probability to C (because its probability could be concentrated at its high bound, which is above C even though its low bound is below C).

Table 9.6 shows an allocation of probabilities to interior cells that minimizes the sum of probabilities in cells whose intervals have high bounds *below* 21 and whose probabilities must therefore contribute to the accumulated probability at $C = 21 - \epsilon$ for small enough ϵ. In this case, the summed probability can be as low as zero, as the table illustrates.

A ↓ $B \rightarrow$	[5, 6] $p = .05$	[6, 7] $p = .06$	[7, 8] $p = .08$	[8, 9] $p = .1$	[9, 10] $p = .21$	[10, 11] $p = .21$	[11, 12] $p = .1$	[12, 13] $p = .08$	[13, 14] $p = .06$	[14, 15] $p = .05$
[10, 11] $p = .1$	[15, 17] $p = \mathbf{0}$	[16, 18] $p = \mathbf{0}$	[17, 19] $p = \mathbf{0}$	[18, 20] $p = \mathbf{0}$	[19, 21] $p = .01$	[20, 22] $p = .01$	[21, 23] $p = 0$	[22, 24] $p = .08$	[23, 25] $p = 0$	[24, 26] $p = 0$
[11, 12] $p = .2$	[16, 18] $p = \mathbf{0}$	[17, 19] $p = \mathbf{0}$	[18, 20] $p = \mathbf{0}$	[19, 21] $p = .1$	[20, 22] $p = 0$	[21, 23] $p = 0$	[22, 24] $p = .1$	[23, 25] $p = 0$	[24, 26] $p = 0$	[25, 27] $p = 0$
[12, 13] $p = .4$	[17, 19] $p = \mathbf{0}$	[18, 20] $p = \mathbf{0}$	[19, 21] $p = 0$	[20, 22] $p = 0$	[21, 23] $p = .2$	[22, 24] $p = .2$	[23, 25] $p = 0$	[24, 26] $p = 0$	[25, 27] $p = 0$	[26, 28] $p = 0$
[13, 14] $p = .2$	[18, 20] $p = \mathbf{0}$	[19, 21] $p = .06$	[20, 22] $p = .08$	[21, 23] $p = 0$	[22, 24] $p = 0$	[23, 25] $p = 0$	[24, 26] $p = 0$	[25, 27] $p = 0$	[26, 28] $p = .06$	[27, 29] $p = 0$
[14, 15] $p = .1$	[19, 21] $p = .05$	[20, 22] $p = 0$	[21, 23] $p = 0$	[22, 24] $p = 0$	[23, 25] $p = 0$	[24, 26] $p = 0$	[25, 27] $p = 0$	[26, 28] $p = 0$	[27, 29] $p = 0$	[28, 30] $p = .05$

Table 9.6. An allocation of probabilities to interior cells that minimizes the accumulated probability at $C = 21 - \epsilon$. The relevant cells are those with intervals whose high bounds are below 21. This comprises 10 cells clustered in the upper left of the table. All of these can contain 0 probability while maintaining consistency with the marginal probabilities for A and B, so the minimum summed probability is zero.

$C = 21$: The minimum cumulative probability at $C = 21$ consists of the minimum possible sum of the probabilities of cells whose intervals have high bounds of 21 or less. This value is 0.05, because the set of cells whose summed probabilities is to be minimized includes the entire first column (which is constrained by the marginal probabilities of B to contain a total probability of 0.05 within its cells), and the table can be arranged so no other probability is allocated within the set of cells in question. Table 9.7 illustrates a way to do this.

A $B \rightarrow$ $\downarrow$	$[5,6]$ $p=.05$	$[6,7]$ $p=.06$	$[7,8]$ $p=.08$	$[8,9]$ $p=.1$	$[9,10]$ $p=.21$	$[10,11]$ $p=.21$	$[11,12]$ $p=.1$	$[12,13]$ $p=.08$	$[13,14]$ $p=.06$	$[14,15]$ $p=.05$
$[10,11]$ $p=.1$	$[15,17]$ $p=$**.05**	$[16,18]$ $p=$**0**	$[17,19]$ $p=$**0**	$[18,20]$ $p=$**0**	$[19,21]$ $p=$**0**	$[20,22]$ $p=0$	$[21,23]$ $p=0$	$[22,24]$ $p=0$	$[23,25]$ $p=0$	$[24,26]$ $p=.05$
$[11,12]$ $p=.2$	$[16,18]$ $p=$**0**	$[17,19]$ $p=$**0**	$[18,20]$ $p=$**0**	$[19,21]$ $p=$**0**	$[20,22]$ $p=.2$	$[21,23]$ $p=0$	$[22,24]$ $p=0$	$[23,25]$ $p=0$	$[24,26]$ $p=0$	$[25,27]$ $p=0$
$[12,13]$ $p=.4$	$[17,19]$ $p=$**0**	$[18,20]$ $p=$**0**	$[19,21]$ $p=$**0**	$[20,22]$ $p=.1$	$[21,23]$ $p=0$	$[22,24]$ $p=.2$	$[23,25]$ $p=.1$	$[24,26]$ $p=0$	$[25,27]$ $p=0$	$[26,28]$ $p=0$
$[13,14]$ $p=.2$	$[18,20]$ $p=$**0**	$[19,21]$ $p=$**0**	$[20,22]$ $p=.04$	$[21,23]$ $p=0$	$[22,24]$ $p=.01$	$[23,25]$ $p=.01$	$[24,26]$ $p=0$	$[25,27]$ $p=.08$	$[26,28]$ $p=.06$	$[27,29]$ $p=0$
$[14,15]$ $p=.1$	$[19,21]$ $p=$**0**	$[20,22]$ $p=.06$	$[21,23]$ $p=.04$	$[22,24]$ $p=0$	$[23,25]$ $p=0$	$[24,26]$ $p=0$	$[25,27]$ $p=0$	$[26,28]$ $p=0$	$[27,29]$ $p=0$	$[28,30]$ $p=0$

Table 9.7. Minimized cumulative probability for $C = 21$. This requires minimizing the summed probabilities of 15 interior cells in the upper left region of the table. These are the cells containing intervals with high bounds at or below 21.

This manual process of minimizing cumulated probability for different values of C (for the right bounding curve) and maximizing it (for the left), if continued, can produce the complete left and right envelopes. However, a method that can be done by computer is desirable. Such a method is described next.

9.8 Finding Points on the Bounding Envelopes with Linear Programming

As the previous section explained, to find the y-axis probability value of a point on the left or right envelope for a given x-axis value, we must maximize or minimize the sum of the probabilities of some subset of the interior cells in a joint distribution tableau. The probabilities in such tables express a kind of discretized joint probability distribution of two random variables. The marginals of these tables constrain how the probabilities of the interior cells can be allocated during the process of maximizing or minimizing a sum of a subset of them. Specifically, the marginal probabilities impose a value on the sum of the probabilities of the interior cells in each *column*, as well as on the sum of the probabilities of the interior cells in each *row*. These marginal values are givens (see tables of previous section).

The preceding paragraph summarizes the need to maximize or minimize a sum given other, constant sums. This type of situation lends itself to **linear programming**, a widely used technique. Numerous software packages, commercial and public domain, exist for solving linear programming problems. Computer program listings for this are even printed in books. Therefore, rather than describe linear programming algorithms, we show how to set up a linear programming problem whose solution is a maximized value giving the y coordinate of a left or right envelope for some x-axis value. We can assume maximization because when the objective is to minimize the summed probabilities of some interior cells, we can simply maximize the sum of the other interior cells, and subtract that value from 1.

The desired linear programming problem consists of the constraints and the sum to be maximized (the objective function, in linear programming terminology). For illustration, consider a simpler joint distribution tableau than the one used in the previous section (Table 9.8).

$Y \in [4,5]$ $p = \frac{1}{4}$	$XY \in [4,10]$ $p =$	$XY \in [8,20]$ $p =$
$Y \in [3,4]$ $p = \frac{1}{2}$	$XY \in [3,8]$ $p =$	$XY \in [6,16]$ $p =$
$Y \in [2,3]$ $p = \frac{1}{4}$	$XY \in [2,6]$ $p =$	$XY \in [4,12]$ $p =$
$Y \Uparrow X \Rightarrow$	$X \in [1,2]$ $p = \frac{1}{2}$	$X \in [2,4]$ $p = \frac{1}{2}$

Table 9.8. A joint distribution tableau showing marginals X and Y and interior cells showing intervals for product XY. Probabilities for the interior cells are left blank because they are not fully determined.

In Table 9.8 the probabilities in the interior cells (which spread out from the northeast corner of the table) are left out because they depend on the dependency relationship between X and Y. Thus, they are variable, although constrained to some degree by the marginal probabilities shown on the left and along the bottom. Linear programming can identify specific probabilities for those interior cells that are (1) consistent with the marginal constraints, and (2) maximize the summed probabilities of any given subset of interior cells.

To solve maximization problems such as these by linear programming, one initializes by assigning feasible values to the variables, which in this case are the interior cell probabilities. These values serve as a starting point from which the linear programming process will automatically find an optimal (maximizing) allocation of probabilities. An initialization method is illustrated next in the joint distribution tableau of Table 9.8.

1. Identify the row with the highest marginal probability, the column with the highest marginal probability, and the interior cell at the intersection of that row and column. The interval in this cell is emphasized in Table

9.9. (In this case, both columns have the same marginal probability, so the first one was chosen.)

$Y \in [4,5]$ $p=\frac{1}{4}$	$XY \in [4,10]$ $p=$	$XY \in [8,20]$ $p=$
$Y \in [3,4]$ $p=\frac{1}{2}$	$XY \in \mathbf{[3,\ 8]}$ $p=$	$XY \in [6,16]$ $p=$
$Y \in [2,3]$ $p=\frac{1}{4}$	$XY \in [2,6]$ $p=$	$XY \in [4,12]$ $p=$
$Y \Uparrow X \Rightarrow$	$X \in [1,2]$ $p=\frac{1}{2}$	$X \in [2,4]$ $p=\frac{1}{2}$

Table 9.9. $XY \in [3,8]$ is chosen as the location for an initial probability assignment.

2. Assign to the identified cell the maximum probability consistent with the row and column marginal constraints affecting it. This is the lesser of the row and column marginal probabilities. See Table 9.10.

$Y \in [4,5]$ $p=\frac{1}{4}$	$XY \in [4,10]$ $p=$	$XY \in [8,20]$ $p=$
$Y \in [3,4]$ $p=\frac{1}{2}$	$XY \in \mathbf{[3,\ 8]}$ $p=\mathbf{1/2}$	$XY \in [6,16]$ $p=$
$Y \in [2,3]$ $p=\frac{1}{4}$	$XY \in [2,6]$ $p=$	$XY \in [4,12]$ $p=$
$Y \Uparrow X \Rightarrow$	$X \in [1,2]$ $p=\frac{1}{2}$	$X \in [2,4]$ $p=\frac{1}{2}$

Table 9.10. An initial probability assignment is made to the interior cell holding interval $[3,8]$.

3. For bookkeeping purposes, subtract the probability just assigned from both corresponding marginal probabilities (Table 9.11).
4. Repeat step 1: Identify a row with the highest marginal probability designation, a column with the highest marginal probability, and the cell at the intersection of that row and column. This cell is the one holding the interval $[8,20]$ in Table 9.11.
5. Repeat step 2: Assign to the newly identified cell the maximum probability consistent with the row and column constraints affecting it. This is the lesser of the row and column probabilities. See Table 9.12.
6. Repeat step 3 in the table most recently modified: To keep track of marginal probability that still needs to be allocated to interior cells, subtract the probability just assigned to an interior cell from the corresponding marginal probabilities (Table 9.13).
7. Repeat step 1 on the current table: Identify a row with the highest amount of as-yet unallocated marginal probability of the rows, a column with the

$Y \in [4,5]$ $p = \frac{1}{4}$	$XY \in [4,10]$ $p =$	$XY \in$**[8, 20]** $p =$
$Y \in [3,4]$ $p' =$**0**	$XY \in [3,8]$ $p = \frac{1}{2}$	$XY \in [6,16]$ $p =$
$Y \in [2,3]$ $p = \frac{1}{4}$	$XY \in [2,6]$ $p =$	$XY \in [4,12]$ $p =$
$Y \Uparrow X \Rightarrow$	$X \in [1,2]$ $p' =$**0**	$X \in [2,4]$ $p = 1/2$

Table 9.11. The probability assigned to the interior cell holding $[3,8]$ is subtracted from the contributing marginal probabilities, which are now labeled p' instead of p to indicate they have been modified. Then the cell holding $[8,20]$ is chosen as the next one to allocate an initial probability to.

$Y \in [4,5]$ $p = \frac{1}{4}$	$XY \in [4,10]$ $p =$	$XY \in$**[8, 20]** $p =$**1/4**
$Y \in [3,4]$ $p' = 0$	$XY \in [3,8]$ $p = \frac{1}{2}$	$XY \in [6,16]$ $p =$
$Y \in [2,3]$ $p = \frac{1}{4}$	$XY \in [2,6]$ $p =$	$XY \in [4,12]$ $p =$
$Y \Uparrow X \Rightarrow$	$X \in [1,2]$ $p' = 0$	$X \in [2,4]$ $p = \frac{1}{2}$

Table 9.12. The cell holding interval $[8,20]$ has its probability assigned a value as high as is consistent with its marginal probabilities.

$Y \in [4,5]$ $p' =$**0**	$XY \in [4,10]$ $p =$	$XY \in [8,20]$ $p =$**1/4**
$Y \in [3,4]$ $p' = 0$	$XY \in [3,8]$ $p = \frac{1}{2}$	$XY \in [6,16]$ $p =$
$Y \in [2,3]$ $p = \frac{1}{4}$	$XY \in [2,6]$ $p =$	$XY \in$**[4, 12]** $p =$
$Y \Uparrow X \Rightarrow$	$X \in [1,2]$ $p' = 0$	$X \in [2,4]$ $p' =$**1/4**

Table 9.13. The allocated probability is subtracted from the relevant marginal cell probabilities, whose remaining unallocated probabilities are designated p'. Then the next cell, holding interval $[4,12]$, is chosen.

highest amount, and the cell at the intersection of that row and column. This cell holds interval $[4,12]$ in Table 9.13.

8. Repeat step 2: Assign to the just-identified cell the maximum probability consistent with the row and column constraints affecting it. This will be the lesser of its row and column unallocated marginal probabilities. See Table 9.14.
9. Repeat step 3: Subtract the initial probability assigned the cell from its corresponding marginal probabilities. See Table 9.15.

$Y \in [4, 5]$ $p' = 0$	$XY \in [4, 10]$ $p =$	$XY \in [8, 20]$ $p = \frac{1}{4}$
$Y \in [3, 4]$ $p' = 0$	$XY \in [3, 8]$ $p = \frac{1}{2}$	$XY \in [6, 16]$ $p =$
$Y \in [2, 3]$ $p = \frac{1}{4}$	$XY \in [2, 6]$ $p =$	$XY \in$**[4, 12]** $p =$**1/4**
$Y \Uparrow X \Rightarrow$	$X \in [1, 2]$ $p' = 0$	$X \in [2, 4]$ $p' = \frac{1}{4}$

Table 9.14. An initial probability is assigned to the cell holding interval $[4, 12]$.

$Y \in [4, 5]$ $p' = 0$	$XY \in [4, 10]$ $p =$	$XY \in [8, 20]$ $p = \frac{1}{4}$
$Y \in [3, 4]$ $p' = 0$	$XY \in [3, 8]$ $p = \frac{1}{2}$	$XY \in [6, 16]$ $p =$
$Y \in [2, 3]$ $p' =$**0**	$XY \in [2, 6]$ $p =$	$XY \in [4, 12]$ $p =$**1/4**
$Y \Uparrow X \Rightarrow$	$X \in [1, 2]$ $p' = 0$	$X \in [2, 4]$ $p'' =$**0**

Table 9.15. The probability assigned to an interior cell is subtracted from the relevant marginals, leaving all zeros in the margins. Key: The number of apostrophes (p, p', or p'') reflects how many times an original marginal probability has been decremented.

10. All marginal probability numbers are now 0, indicating that no marginal probability remains to be allocated to interior cells. Therefore, any interior cells not yet assigned an initial probability must be assigned 0 (Table 9.16).

$Y \in [4, 5]$ $p' = 0$	$XY \in [4, 10]$ $p =$**0**	$XY \in [8, 20]$ $p = \frac{1}{4}$
$Y \in [3, 4]$ $p' = 0$	$XY \in [3, 8]$ $p = \frac{1}{2}$	$XY \in [6, 16]$ $p =$**0**
$Y \in [2, 3]$ $p' = 0$	$XY \in [2, 6]$ $p =$**0**	$XY \in [4, 12]$ $p = \frac{1}{4}$
$Y \Uparrow X \Rightarrow$	$X \in [1, 2]$ $p' = 0$	$X \in [2, 4]$ $p'' = 0$

Table 9.16. The interior cells are now fully initialized and ready for a linear programming process to modify them to an optimal set of assignments that maximizes.

11. Since the interior cells are now initialized appropriately, the marginal probability designations used for bookkeeping purposes are no longer needed and can be replaced with the actual marginal probability values that were originally present. This results in Table 9.17, which serves as input to the linear programming problem.

$Y \in [4,5]$ $p = \frac{1}{4}$	$XY \in [4,10]$ $p_{11} = 0$	$XY \in [8,20]$ $p_{21} = \frac{1}{4}$
$Y \in [3,4]$ $p = \frac{1}{2}$	$XY \in [3,8]$ $p_{12} = \frac{1}{2}$	$XY \in [6,16]$ $p_{22} = \frac{1}{4}$
$Y \in [2,3]$ $p = \frac{1}{4}$	$XY \in [2,6]$ $p_{13} = 0$	$XY \in [4,12]$ $p_{23} = \frac{1}{4}$
$Y \Uparrow X \Rightarrow$	$X \in [1,2]$ $p = \frac{1}{2}$	$X \in [2,4]$ $p = \frac{1}{2}$

Table 9.17. The interior cells have been initialized and the table is ready for linear programming to be applied.

Table 9.17 also gives distinctive subscripts to the interior cell probabilities so that they can be referred to individually. From this table, linear programming will find the best allocation of marginal probabilities over interior cells, which is the one with the maximum value possible for the sum of the probabilities of a designated subset of the interior cells. The linear programming problem takes as input all of the row and column constraints plus the optimization (or "objective") function, which is the sum of the interior cell probabilities to be maximized. For Table 9.17, this input is shown in Table 9.18.

Value	p_{11}	p_{21}	p_{12}	p_{22}	p_{13}	p_{23}
1/4	1	1	0	0	0	0
1/2	0	0	1	1	0	0
1/4	0	0	0	0	1	1
1/2	1	0	1	0	1	0
1/2	0	1	0	1	0	1
3/4	1	0	1	0	1	1

Table 9.18. The five constraints (three row + two column) from Table 9.17 are shown in the first five rows, followed in the last row by the optimization function for maximizing the sum of the probabilities of the four interior cells whose interval low bound is below 6. The 1's and 0's are coefficients of the p_{ij} from Table 9.17.

The linear programming process will take a chart like Table 9.18 and find values for the various p_{ij} that maximize the number at the bottom of the Value column. In the case shown, this is initially 3/4, the sum of the probabilities whose associated intervals have low bounds below 6. Note that the initial values of the probabilities shown in Table 9.17 determine the initial value of 3/4 for the optimization function. The 1's in Table 9.18 are coefficients that designate which probabilities are governed by which constraints. Thus, the first row says that $1 * p_{11} + 1 * p_{21} + 0 * p_{12} + 0 * p_{22} + 0 * p_{13} + 0 * p_{23} = \frac{1}{4}$ or, equivalently, $p_{11} + p_{21} = \frac{1}{4}$. This is a constraint stated by Table 9.17, as are the next four rows. The remaining, last row is not a constraint but rather the optimization equation. Hence, the value of 3/4 is not fixed, as it would be for a constraint, but can vary. Normally it does vary, as the linear programming algorithm tries to maximize it.

9.9 Conclusion

An introduction to the DEnv approach has been presented at the tutorial level. More advanced features are available, and a considerable amount of related work by others has appeared. Regarding advanced features, one is the use of correlation between two random variables to supplement the basic row and column constraints imposed by a joint distribution tableau. This is described in detail in [5]. A slightly less general, but more accessible, discussion of correlation along with an application to reliability of two-component systems appears in [6]. Many joint distributions encountered in practice are unimodal. Unimodality constraints in the DEnv approach and its software implementation are discussed in [18]. A more theoretical discussion from the Kreinovich lab appears in [7]. A tool and related documentation is available for download at http://ifsc.ualr.edu/jdberleant/statool/index.htm.

Work on bounded families of distributions has experienced a surge of interest in recent years. Considerable work has appeared in the biannual International Symposium on Imprecise Probabilities: Theories and Applications (ISIPTA) sponsored by the eponymous society (http://sipta.org/). The focus of the biannual Workshop on Reliable Engineering Computing (www.gtsav.gatech.edu/workshop/rec08, .../rec06, and .../rec04) is even more apropos. The most closely related and coherent compendium of work is still Helton and Oberkampf [12], a collection of papers all focusing on the same set of challenge problems concerning system response under uncertainty. Various alternatives to the DEnv algorithm are explored in the context of this set of problems. Also closely related are reports by Ferson et al. [9, 10]. Although well-known in the field, Kuznetsov [14] is unfortunately currently unavailable in English. Other books of interest include [8, 11, 15, 16]). All of these works deal, as does the present work, with the important problem of drawing what conclusions are possible in the presence of incomplete information: inference under severe uncertainty.

References

1. Berleant, D., Anderson, G.T.: Decision-making under severe uncertainty for autonomous mobile robots. In: Proceedings of the IEEE International Conference on Systems, Man and Cybernetics (2007)
2. Berleant, D., Cheng, H.: A software tool for automatically verified operations on intervals and probability distributions. Reliable Computing 4, 71–82 (1998)

3. Berleant, D., Goodman-Strauss, C.: Bounding the results of arithmetic operations on random variables of unknown dependency using intervals. Reliable Computing 4, 147–165 (1998)
4. Berleant, D., Zhang, J.: Representation and problem solving with distribution envelope determination (DEnv). Reliability Engineering and System Safety 85, 153–168 (2004)
5. Berleant, D., Zhang, J.: Using Pearson correlation to improve envelopes around the distributions of function. Reliable Computing 10, 139–161 (2004)
6. Berleant, D., Zhang, J.: Bounding the times to failure of 2-component systems. IEEE Transactions on Reliability 53, 542–550 (2004)
7. Berleant, D., Kosheleva, O., Kreinovich, V., Nguyen, H.T.: Unimodality, independence lead to NP–hardness of interval probability problems. Reliable Computing 13, 261–282 (2007)
8. Fellin, W., Lessmann, H., Oberguggenberger, M., Vieider, R.: Analyzing Uncertainty in Civil Engineering. Springer-Verlag, Berlin (2005)
9. Ferson, S., Hajagos, J., Berleant, D., Zhang, J., Tucker, W.T., Ginzburg, L., Oberkampf, W.: Dependence in Dempster–Shafer theory and probability bounds analysis. Tech. rep., Sandia National Laboratory (2004). Report SAND2004-3072
10. Ferson, S., Kreinovich, V., Ginzburg, L., Myers, D.S., Sentz, K.: Constructing probability boxes and Dempster–Shafer structures. Technical Report SAND2002-4015, Sandia National Laboratories (2003).
11. Halpern, J.Y.: Reasoning about uncertainty. MIT Press, Cambridge, MA (2003)
12. Helton, J.C., Oberkampf, W.L.: Special issue on alternative representations of epistemic uncertainty. Reliability Engineering and System Safety 85, 1–369 (2004)
13. Kolmogoroff, A.: Confidence limits for an unknown distribution function. Annals of Mathematical Statistics 12, 461–463 (1941)
14. Kuznetsov, V.: Interval statistical models. Radio i Svyaz, Moscow (1991)(in Russian)
15. Manski, C.: Partial Identification of Probability Distributions. Springer-Verlag, New York (2003)
16. Walley, P.: Statistical Reasoning with Imprecise Probabilities. Chapman & Hall, New York (1990)
17. Wang, R., Pierce, E., Madnick, S., Fisher, C. (eds.): Information Quality. M. E. Sharpe, Armonk, NY (2005)
18. Zhang, J., Berleant, D.: Arithmetic on random variables: squeezing the envelopes with new joint distribution constraints. In: Proceedings of the Fourth International Symposium on Imprecise Probabilities and Their Applications (ISIPTA '05), pp. 416–422 (2005)

10

IntBox: An Object-Oriented Interval Computing Software Toolbox in C++

Michael Nooner and Chenyi Hu

Computer Science Department, University of Central Arkansas, Conway, AR 72035-0001, USA. mnooner,chu@uca.edu

In this book we have discussed applying interval methods in knowledge processing. To effectively realize these methods, an interval computing environment is needed. In this chapter, we present an object-oriented interval software toolbox written in C++. This toolbox, named *IntBox*,[1] supports algebraic, utility, and set operations among intervals, interval vectors, and interval matrices. The design and implementation of this interval software toolbox follow recently proposed interval computing standards. The toolbox can be easily embedded into an ANSI/ISO C++ environment to enable interval software development. It is portable, easy to use, well tested, and robust with built-in error handling features. Instructions on package installation, testing, usage, and sample applications are included.

10.1 Introduction

There are numerous software tools for interval computing written in mainstream languages such as C, C++, Fortran, Java, as well as in computational algebra systems, such as Maple, MATLAB, and Mathematica. However, these packages are mostly designed and implemented for specific systems. Portable software tools and techniques in compliance with established standards are very helpful for both software developers and users. Recent efforts to standardize interval computing [3], [2] make it possible to develop a generalized portable interval software development toolbox.

In the design of this toolbox, we selected the modular architecture emphasized in object-oriented programming. In interval computing, the operands involved include intervals, interval vectors, and/or interval matrices. It is essential to instantiate objects of interval, interval vector, and interval matrix types. Therefore, three basic classes: - `Interval`, `IntervalVector`, and

[1] The complete package is available at `http://www.cs.uca.edu/interval/intbox`. An earlier version was initially reported in [5].

C. Hu et al. (eds.), *Knowledge Processing with Interval and Soft Computing*, DOI: 10.1007/978-1-84800-326-2_10,

`IntervalMatrix` - are included. Using the polymorphism and encapsulation features of C++, this toolbox allows operations involving interval objects and standard C++ objects without extra efforts from the programmer.

10.2 The Interval Class

10.2.1 Intervals

A nonempty *mathematical interval* $[a, b]$ is the set $\{x \in \mathbb{R} \mid |a \leq x \leq b\}$, where $a \leq b$. A *machine interval* $[a^*, b^*]$ is a mathematical interval whose endpoints are machine-representable numbers. We say that $[a^*, b^*]$ is a machine representation of $[a, b]$ if $[a^*, b^*]$ *contains* $[a, b]$ (i.e., $a^* \leq a$ and $b \leq b^*$). We say that the machine interval $[a^*, b^*]$ is a *sharp representation* of a mathematical interval $[a, b]$ if and only if a^* is the greatest machine-representable number that is less than or equal to a; and b^* is the least machine-representable number that is greater than or equal to b. The *empty interval* $\emptyset$, which does not contain any real number, is required in this toolbox.

Interval arithmetic on mathematical intervals is defined as follows:

> Let $\boldsymbol{a}$ and $\boldsymbol{b}$ be two mathematical intervals. Let $\mathtt{op}$ be one of the arithmetic operations $+, -, \times$, or $\div$. Then $\boldsymbol{a}\ \mathtt{op}\ \boldsymbol{b} \equiv \{a\ \mathtt{op}\ b \mid a \in \boldsymbol{a}, b \in \boldsymbol{b}\}$. It is an exception[2] if $\mathtt{op}$ represents $\div$ and $0 \in \boldsymbol{b}$. If either $\boldsymbol{a}$ or $\boldsymbol{b}$ is empty, then $\boldsymbol{a}$ op $\boldsymbol{b}$ is the empty interval.

All operations inside a computer are performed on machine intervals. Arithmetic on machine intervals must satisfy the following condition:

> *Containment Constraint:* Let $\boldsymbol{a} = [\underline{a}, \overline{a}]$ and $\boldsymbol{b} = [\underline{b}, \overline{b}]$ be intervals. Let $\boldsymbol{c} = [\underline{c}, \overline{c}]$ be the interval result of computing $\boldsymbol{a}\ \mathtt{op}\ \boldsymbol{b}$. Then $\boldsymbol{c}$ must contain the exact mathematical interval $\boldsymbol{a}\ \mathtt{op}\ \boldsymbol{b}$.

In other words, interval arithmetic on nonempty machine intervals requires that we round down the lower bound and round up the upper bound, if they are not machine representable, to guarantee that the machine interval result contains the true mathematical interval result. This is needed to produce guaranteed error bounds.

10.2.2 The Boost Library's Interval Class

In late 2006, the C++ standardization committee evaluated and responded positively to the proposal to add Interval Arithmetic as a part of ANSI/ISO C++ Standard Library [3]. In the proposal, the peer-reviewed Boost interval

---

[2] There has been significant progress on standardizing extended interval arithmetic (involving division by intervals that contain zero) recently. For example, see [6].

arithmetic library, `interval_lib` [4], is often cited as a prototype implementation of the C++ interval arithmetic standard. It is our belief that Boost's interval library will be very similar to the approved standard. Therefore, we chose to use the Boost interval library for fundamental interval operations in IntBox. Boost's implementation was also chosen for two other reasons. First, the Boost license is very flexible in that we are able to redistribute just the `interval_lib` portion of the much larger Boost package. Second, the Boost library is highly cross-platform and cross-compiler compatible, which is an important aspect of the design objectives.

The Boost package uses the class `interval` to perform interval operations. The Boost interval class uses zero for the default value of an interval. Also by default, the `interval` class dislikes empty intervals. In fact, it will throw an exception if one is created. However, interval set operations may result in the empty interval as a valid output. Therefore, proper handling of empty intervals is required. Boost's interval library does allow the user to set various policies, including not throwing exceptions because of the empty intervals. We created a template specialization called `defualt_interval` that sets such a policy.

10.2.3 Functionality for the Interval Class

The functionality for the `interval` class is grouped into the following categories: arithmetic, set, logic, and utility. Let $\boldsymbol{a}$ and $\boldsymbol{b}$ be two intervals. We list implementational definitions for each of these operations associated with the operator implemented in this toolbox.

- Arithmetic operations:
  - Addition (operator $+$): $\boldsymbol{a}+\boldsymbol{b} = [\underline{a}+\underline{b}, \overline{a}+\overline{b}]$ if neither $\boldsymbol{a}$ nor $\boldsymbol{b}$ is empty; otherwise, $\boldsymbol{a}+\boldsymbol{b} = \emptyset$.
  - Subtraction (operator $-$): $\boldsymbol{a}-\boldsymbol{b} = [\underline{a}-\overline{b}, \overline{a}-\underline{b}]$ if neither $\boldsymbol{a}$ nor $\boldsymbol{b}$ is empty; otherwise, $\boldsymbol{a}+\boldsymbol{b} = \emptyset$.
  - Multiplication (operator $*$): $\boldsymbol{a}*\boldsymbol{b} = [\min\{\underline{a}\underline{b}, \underline{a}\overline{b}, \overline{a}\underline{b}, \overline{a}\overline{b}\}, \max\{\underline{a}\underline{b}, \underline{a}\overline{b}, \overline{a}\underline{b}, \overline{a}\overline{b}\}]$ if neither $\boldsymbol{a}$ nor $\boldsymbol{b}$ is empty; otherwise, $\boldsymbol{a}*\boldsymbol{b} = \emptyset$.
  - Division (operator $/$): $\boldsymbol{a}/\boldsymbol{b} = [\min\{\underline{a}/\underline{b}, \underline{a}/\overline{b}, \overline{a}/\underline{b}, \overline{a}/\overline{b}\}, \max\{\underline{a}/\underline{b}, \underline{a}/\overline{b}, \overline{a}/\underline{b}, \overline{a}/\overline{b}\}]$ if neither $\boldsymbol{a}$ nor $\boldsymbol{b}$ is empty and $0 \notin \boldsymbol{b}$; if $\boldsymbol{a}$ or $\boldsymbol{b}$ is empty, $\boldsymbol{a}/\boldsymbol{b} = \emptyset$; otherwise, if $0 \in \boldsymbol{b}$ an exception is thrown.[3]
  - Cancellation (method `cancel()`): $\boldsymbol{a}$.`cancel`$(\boldsymbol{b}) = [\underline{a}-\underline{b}, \overline{a}-\overline{b}]$ if neither $\boldsymbol{a}$ nor $\boldsymbol{b}$ is empty and $\overline{a}-\underline{a} \geq \overline{b}-\underline{b}$; otherwise, $\boldsymbol{a}$.`cancel`$(\boldsymbol{b}) = \emptyset$.
- Set operations among intervals.
  - Intersection (operator &): If $\max\{\underline{a}, \underline{b}\} \leq \min\{\overline{a}, \overline{b}\}$ $\boldsymbol{a} \cap \boldsymbol{b} = [\max\{\underline{a}, \underline{b}\}, \min\{\overline{a}, \overline{b}\}]$; otherwise, $\boldsymbol{a} \cap \boldsymbol{b} = \emptyset$.

[3] The package may be extended to allow division by intervals that contain zero by adopting an implementation of extended interval arithmetic standard.

- Hull (method `hull()`): If both $\boldsymbol{a}$ and $\boldsymbol{b}$ are empty, then hull of $\boldsymbol{a}$ and $\boldsymbol{b}$ is empty; if only one of $\boldsymbol{a}$ and $\boldsymbol{b}$ is empty, then hull of $\boldsymbol{a}$ and $\boldsymbol{b}$ is the same as the nonempty interval; otherwise, then hull of $\boldsymbol{a}$ and $\boldsymbol{b}$ is $[\min\{\underline{a}, \underline{b}\}, \max\{\overline{a}, \overline{b}\}]$.

- Logical operations:
  - Equality (operator (==)): If $\underline{a} = \underline{b}$ and $\overline{b} = \overline{a}$, then $\boldsymbol{a} = \boldsymbol{b}$.
  - Less than[4] (operator (<)): If $\overline{a} < \underline{b}$, then $\boldsymbol{a} < \boldsymbol{b}$.
  - Greater than operator >): If $\underline{a} > \overline{b}$, then $\boldsymbol{a} > \boldsymbol{b}$.
  - Disjoint test (method `disjoint()`): If $\boldsymbol{a} \cap \boldsymbol{b} = \emptyset$, `disjoint(`$\boldsymbol{a}$`, `$\boldsymbol{b}$`)` returns true.
  - Interior test (method `interior()`: If $(\boldsymbol{b} \subset \boldsymbol{a})$, then $\boldsymbol{b}$ is an interior of $\boldsymbol{a}$.
- Utility functions:
  - Midpoint (method `midpoint()`: The midpoint of an interval $\boldsymbol{a}$ is $(\underline{a} + \overline{a})/2$ provided $\boldsymbol{a}$ is not empty.
  - Width (method `width()`: The width an interval $\boldsymbol{a}$ is $\overline{a} - \underline{a}$ provided $\boldsymbol{a}$ is not empty.
  - Assignment (operator =): Stores the value of an interval to the left operand.
  - Output (operator <<): Stream insertion operator for output.
  - Input (operator >>): Stream extraction operator for input.

In addition to above, the class supports elementary functions for intervals. These include interval power, square root, exponent, logarithm, trigonometry, and inverse trigonometry functions. Let f be an elementary function and let $\boldsymbol{x}$ be an interval. Then $f(\boldsymbol{x})$ returns an interval that contains $f(x)\ \forall x \in \boldsymbol{x}$.

10.2.4 Sample Code

Here is a segment of example code that defines and accesses interval objects; more specifically the `Interval` specialization is used. The code defines three interval objects: `a, b`, and `c` with the default, single, and double parameters, respectively.

The statement `c < b` returns true. Hence, the statement that outputs the product of the two intervals is executed. The next statements find the tangent of the interval `c` and then store the results in in the interval `a`.

```
default_interval a, b = 3.0, c(0.1, 0.2);
if (c < b)
   cout << "c * b = " << c * b << endl;
a = tan(c);
cout << "The interval containing tangent "
     << c << " is: \n" << a << endl;
```

[4] The operators <, and > check only the crisp relation.

The above code segment outputs:

```
c * b = [0.3 ; 0.6]
The interval containing tangent [0.1 ; 0.2] is:
[0.100335 ; 0.20271]
```

In the output, we use a semicolon to separate the lower and upper bounds of an interval, as suggested by the proposed C++ interval standard library.

10.3 Functionalities Involving Interval Vector and Matrices

Similar to floating point linear algebra, interval linear algebra is useful in most calculations. The Basic Linear Algebra Subprograms Technical Forum [1] has established an updated BLAS standard [2], in which a proposed interval BLAS (iBLAS) standard was included in the journal of development (available at `http://www.netlib.org/blas/blast-forum/chapter5.pdf`). As in iBLAS, we use uppercase boldface letters, like $\boldsymbol{A}, \boldsymbol{B}$, and $\boldsymbol{C}$, to denote interval matrices; lowercase boldface letters are used for interval vectors; boldface Greek letters denote interval scalars. The transpose of an interval matrix $\boldsymbol{A}$ is denoted by $\boldsymbol{A}^T$.

We follow the iBLAS standard in the design and implementation of this toolbox. However, by applying features of the object-oriented programming paradigm, this toolbox contains much more functionality than that of the iBLAS. The functionalities related to interval linear algebra are grouped into interval vector operations, interval matrix-vector operations, interval matrix-matrix operations, and set and utility operations for interval vectors and interval matrices.

10.3.1 Functionality Involving Interval Vectors

We list the main functionalities involving interval vector operations in this toolbox.

- Interval vector reduction operations:
  - Dot product (method `dot`): $\boldsymbol{r} \leftarrow \boldsymbol{\beta r} + \boldsymbol{\alpha x}^T \boldsymbol{y}$
  - Vector norms (method `norm`): $r \leftarrow ||\boldsymbol{x}||_1, ||\boldsymbol{x}||_2, ||\boldsymbol{x}||_\infty$
  - Sum (method `sum`): $\boldsymbol{r} \leftarrow \sum_i \boldsymbol{x}_i$
  - Max magnitude and location (method `amax_val`): $k, \boldsymbol{x}_k$; $k =$ arg, $\max_i\{|\underline{x}_i|, |\overline{x}_i|\}$
  - Min absolute value and location (method `amin_val`): $k, \boldsymbol{x}_k$; $k =$ arg, $\min_i\{|\underline{x}_i|, |\overline{x}_i|\}$ if $0 \notin \boldsymbol{x}_i \forall i$; 0 otherwise
  - Sum of squares (method `sumsq`): $(\boldsymbol{a}, \boldsymbol{b}) \leftarrow \sum_i \boldsymbol{x}_i^2$, $\boldsymbol{a} \cdot \boldsymbol{b}^2 = \sum_i \boldsymbol{x}_i^2$
- Interval vector operations:

- Addition/subtraction (operator $+/-$): elementwise addition or subtraction of two interval vectors with the same length
- Scale an interval vector (operator $*$): $\boldsymbol{\alpha} * \boldsymbol{x}$, where $\boldsymbol{\alpha}$ is an interval.
- Reciprocal scale (method `rscale`): $\boldsymbol{x} \leftarrow \boldsymbol{x}/\boldsymbol{\alpha}$
- Scaled interval vector accumulation (method `axpby`): $\boldsymbol{y} \leftarrow \boldsymbol{\alpha x} + \boldsymbol{\beta y}$
- Scaled interval vector accumulation (method `waxpby`): $\boldsymbol{w} \leftarrow \boldsymbol{\alpha x} + \boldsymbol{\beta y}$
- Scaled interval vector cancellation (method `cancel`): $\boldsymbol{y} \leftarrow \boldsymbol{\alpha x} \ominus \boldsymbol{\beta y}$
- Scaled interval vector cancellation (method `wcancel`): $\boldsymbol{w} \leftarrow \boldsymbol{\alpha x} \ominus \boldsymbol{\beta y}$

• Data movement with interval vector:
  - Copy (operator $=$, or method `copy`): $\boldsymbol{y} \leftarrow \boldsymbol{x}$
  - Swap (method `swap`): $\boldsymbol{y} \leftrightarrow \boldsymbol{x}$
  - Permute vector (method `permute`): $\boldsymbol{x} \leftarrow P\boldsymbol{x}$

• Interval vector-matrix operations:
  - Matrix vector product (operator $*$ or method `mv`): $\boldsymbol{A}*\boldsymbol{x}$ or $\boldsymbol{y} \leftarrow \boldsymbol{\alpha A x} + \boldsymbol{\beta y}$, or $\boldsymbol{y} \leftarrow \boldsymbol{\alpha A}^T\boldsymbol{x} + \boldsymbol{\beta y}$, or $\boldsymbol{x} \leftarrow \boldsymbol{T x}, \boldsymbol{x} \leftarrow \boldsymbol{T}^T\boldsymbol{x}$
  - Rank one updates (method `rv`): $\boldsymbol{A} \leftarrow \boldsymbol{\alpha x y}^T + \boldsymbol{\beta A}$

• Set operations involving interval vectors:
  - Enclosed (method `encv`): $\boldsymbol{x}$ is enclosed in $\boldsymbol{y}$ if $\boldsymbol{x} \subseteq \boldsymbol{y}$.
  - Interior (method `interiorv`): $\boldsymbol{x}$ is enclosed in the interior of $\boldsymbol{y}$.
  - Disjoint (method `disjv`): $\boldsymbol{x}$ and $\boldsymbol{y}$ are disjoint if $\boldsymbol{x} \cap \boldsymbol{y} = \emptyset$.
  - Intersection (method `interv`): $\boldsymbol{y} \leftarrow \boldsymbol{x} \cap \boldsymbol{y}, \boldsymbol{z} \leftarrow \boldsymbol{x} \cap \boldsymbol{y}$.
  - Hull (method `hullv`): the convex hull of $\boldsymbol{x}$ and $\boldsymbol{y}$.

• Utility operations for interval vectors:
  - Empty element (method `emptyelev`): k if $\boldsymbol{x}_k = \emptyset$; or -1
  - Left endpoint (method `infv`): $v \leftarrow \underline{x}$
  - Right endpoint (method `supv`): $v \leftarrow \overline{x}$
  - Midpoint (method `midv`): $v \leftarrow (\underline{x} + \overline{x})/2$
  - Width (method `widthv`): $v \leftarrow \overline{x} - \underline{x}$
  - Construct (method `constructv`): $\boldsymbol{x} \leftarrow u, v$
  - Insertion (operator <<): for output an interval vector
  - Equality test (operator ==): if two interval vectors are equal

10.3.2 Functionality Involving Interval Matrices

We list the main functionalities of this toolbox involving interval matrix operations. As specified in the iBLAS standard, matrices involved can be general, symmetric, triangular, general band, symmetric band, triangular band, and so on.

• $O(n^2)$ matrix operations:
  - Matrix norms (method `norm`): $r \leftarrow ||\boldsymbol{A}||_1, ||\boldsymbol{A}||_{F,}, ||\boldsymbol{A}||_\infty, ||\boldsymbol{A}||_{\max}$
  - Diagonal scaling (method `diag_scale`): $\boldsymbol{A} \leftarrow \boldsymbol{DA}, \boldsymbol{AD}$
  - Two sided di-scaling (method `lrscale`): $\boldsymbol{A} \leftarrow \boldsymbol{D}_1\boldsymbol{A}\boldsymbol{D}_2$, $\boldsymbol{DAD}$, $\boldsymbol{A} + \boldsymbol{BD}$

  - Matrix accumulation and scale (method acc): $\boldsymbol{B} \leftarrow \alpha\boldsymbol{A}+\beta\boldsymbol{B}, \alpha\boldsymbol{A}^T+\beta\boldsymbol{B}$
  - Matrix add and scale (method add): $\boldsymbol{C} \leftarrow \alpha\boldsymbol{A}+\beta\boldsymbol{B}$
- Matrix matrix product (operator $*$, or method mm): $\boldsymbol{A}*\boldsymbol{B}$, or $\boldsymbol{C} \leftarrow \alpha\boldsymbol{A}\boldsymbol{B}+\beta\boldsymbol{C}, \alpha\boldsymbol{A}^T\boldsymbol{B}+\beta\boldsymbol{C}, \alpha\boldsymbol{A}\boldsymbol{B}^T+\beta\boldsymbol{C}, \alpha\boldsymbol{A}^T\boldsymbol{B}^T+\beta\boldsymbol{C}, \alpha\boldsymbol{B}\boldsymbol{A}+\beta\boldsymbol{C}, \alpha\boldsymbol{B}^T\boldsymbol{A}+\beta\boldsymbol{C}, \alpha\boldsymbol{B}\boldsymbol{A}^T+\beta\boldsymbol{C}, \alpha\boldsymbol{B}^T\boldsymbol{A}^T+\beta\boldsymbol{C}$
- Data movement with interval matrices:
  - Matrix copy (operator =, or method copy): $\boldsymbol{B} \leftarrow \boldsymbol{A}, \boldsymbol{A}^T$
  - Matrix transpose (method trans): $\boldsymbol{A} \leftarrow \boldsymbol{A}^T$
  - Permute matrix (method permute): $\boldsymbol{A} \leftarrow \boldsymbol{P}\boldsymbol{A}, \boldsymbol{A}\boldsymbol{P}$
- Set operations for interval matrices:
  - Enclosed (method encm): $\boldsymbol{A}$ is enclosed in $\boldsymbol{B}$ if $\boldsymbol{A} \subseteq \boldsymbol{B}$.
  - Interior (method interiorm): $\boldsymbol{A}$ is enclosed in the interior of $\boldsymbol{B}$.
  - Disjoint (method disjm): $\boldsymbol{A}$ and $\boldsymbol{B}$ are disjoint if $\boldsymbol{A} \cap \boldsymbol{B} = \emptyset$.
  - Intersection (method interm): $\boldsymbol{B} \leftarrow \boldsymbol{A} \cap \boldsymbol{B}, \boldsymbol{C} \leftarrow \boldsymbol{A} \cap \boldsymbol{B}$.
  - Hull (method hullm): the convex hull of $\boldsymbol{A}$ and $\boldsymbol{B}$.
- Utility operations for interval matrices:
  - Empty element (method emptyelem): if $\boldsymbol{A}_{ij} = \emptyset$ for some i and j
  - Left endpoint (method infm): $C \leftarrow \underline{A}$
  - Right endpoint (method supm): $C \leftarrow \overline{A}$
  - Midpoint (method midm): $C \leftarrow (\underline{A}+\overline{A})/2$
  - Width (method widthm): $C \leftarrow \overline{A}-\underline{A}$
  - Construct (method constructm): $\boldsymbol{A} \leftarrow B, C$
  - Insertion (operator <<): for output an interval matrix
  - Equality test (operator ==): if two interval matrices are equal

We implemented the above iBLAS functionalities in two C++ template classes `IntervalVector <I>` and `IntervalMatrix <I>`. The template parameter (i.e., `I`), expects a Boost `interval<T,P>` type. By default, for the reasons discussed earlier, the `default_interval` specialization is used. This object-oriented approach provides this toolbox with many user-friendly build-in features.

10.4 The `IntervalMatrix` Class

The `IntervalMatrix<I>` class is for a general matrix and is the topmost base class of all other structured matrix classes. Hence, it defines the general properties that are shared by all of the more specialized matrices: symmetric, triangular, general band, symmetric band, triangular band, and so on. The underlying interval elements of the class are stored in a one-dimensional partitioned array, consistently with the BLAS standard. The sample code at the end of this section provides an example of using this base class. Detailed documentation and source code for this class are included in the package.

10.4.1 Creation

There are several different constructors. The most commonly used constructor simply takes the dimensions of the matrix as two parameters. Another constructor creates an interval matrix from two arrays, where the first and the second arrays are the lower and upper bounds of an interval matrix, respectively. For robustness, an element in the first array does not necessarily have to be less than the corresponding element of the second; the two bounds will simply be reversed. It is the user's responsibility, if needed, to ensure that the input lower bound is actually less than or equal to the upper bound. Finally, there is also a default constructor, which creates a matrix that has no underlying array and is of size 0×0. The usage of the default constructor is described in the following subsection. The size of a matrix is set by the above three constructors and is fixed. The size of a matrix can be retrieved using the `getRows()` and `getCols()` methods.

10.4.2 Assignment

As mentioned earlier, the dimensions of a matrix cannot be resized. There is one exception to this rule, and that is assignment. For robustness, the assignment operation disregards dimensions; it will resize the underlying array as needed. Therefore, the default constructor is useful in that it allows one to create unsized matrices that have no underlying array. Such matrices are suitable for catching results.

The assignment operator uses the following rules to determine when to allocate a new array. If the source matrix is smaller than or the same size as the destination, then the array is not resized. Furthermore, if the right-hand argument is an intermediate result (discussed in Section 10.4.5) and has the same storage structure, then only a pointer assignment is needed, rather than an elementwise copy.

10.4.3 Accessing Elements

The elements of an interval matrix can be accessed three ways. First, there are the `getAt()` and `setAt()` methods. These methods allow getting and setting of a given element in the matrix. Importantly, they check that the given coordinates are within the bounds of the matrix. Alternatively, elements within the matrix can be accessed by two other methods called `getAtNC()` and `setAtNC()`. These two methods function like `getAt()` and `setAt()` except that no dimension checking is done. These methods are used extensively internally, because they add a modest increase to efficiency, by reducing the amount of checking. Finally, there are two operators that allow access to the elements of a matrix. The bracket operator (i.e., [ ]), has been overloaded in such a way that the class can be treated similarly to a two-dimensional array. Also, the ( ) operator has been overloaded to take two parameters (i.e., a row and column index). Both operators simply offer a convenient wrapper for the `getAt()` method.

10.4.4 Elementary Arithmetic Operations and Boolean Methods

Elementary arithmetic operations (e.g. addition and subtraction), are implemented by overloading these operators: `+` (addition), `-` (subtraction), and `*` (multiplication). Each of these operators, whenever appropriate, is valid for matrix-matrix, matrix-vector, vector-matrix, matrix-interval, and interval-matrix operations. When using pointers, utilizing overloaded operators can be troublesome. Hence, each of these operators has an associated method that performs the operation (i.e. `add()`, `sub()`, and `mult()`).

It is often useful to compare two matrices. With this in mind, several comparison methods are offered. First, the `==` (equals) and `!=` (not equals) operators have been overloaded. Two matrices are considered equal if the corresponding elements in both matrices are equal. The method `isEmpty()` will return true if the matrix contains an empty element. The method `isSymmetric()` can be used to tell whether the matrix is symmetric. The method `isTriangular()` will return true if the matrix has a triangular shape. The proposed iBLAS standard defines three set tests that a library must have. First is the `enc()` method, which tests to see if the elements of a matrix are supersets of the corresponding elements of another matrix. Second is the `interior()` method, which acts like `enc()` except the elements are tested to see if the elements of another matrix are proper subsets. Finally, the `disj()` method returns true if all of the corresponding elements in two matrices are disjoint.[5]

10.4.5 iBLAS Methods

The iBLAS functionalities are implemented as methods. The parameters of an iBLAS method are structured as follows: first matrices or vectors, then transpose flags, next a side flag, and, finally, any scalars. For example, below is the function header for the `mm()` method, which solves formulas like $\alpha \mathbf{A}^{\mathrm{T}} * \mathbf{B}^{\mathrm{T}} + \beta \mathbf{C}$.

```
IntervalMatrix<I>& mm( IntervalMatrix<I>& b, //Matrix
                IntervalMatrix<I>& c,        //Matrix
                bool transpose_this_matrix,  //Transpose flag
                bool b_transpose,            //Transpose flag
                bool on_left,                //Side flag
                const I& alpha,              //Scalar
                const I& beta,               //Scalar
                IntervalMatrix<I>& result_mat )
```

[5] Although this is not necessary mathematically, the implementation ensures that there is no overlap in any projected subspaces.

The calling object takes the place of the matrix **A** in the formula with the parameters given above. This is a general rule for all the iBLAS functions. Notice that the above example returns the result of the calculation. What is actually returned is either a recycled parameter or a new intermediate result object allocated off the heap space. This can be very inefficient, especially for large matrices. However, returning a value is useful when mixing operators and method calls together. To address these concerns, the `result_mat` parameter can be used. This is an *optional* parameter that always appears last in every method. If a matrix is specified, then that matrix will be used instead of a new or recycled matrix. Be aware that `result_mat` must be of the proper size to hold the results of the calculation; if it is not, then an exception will be thrown; that is, it will not be resized. Furthermore, if the `result_mat` parameter is specified, then a reference to `result_mat` is the returned value.

When the side flag is true, then the calling object is on the left side of the multiplication, and when it is false the calling object is on the right side of the multiplication (e.g., $B * A$). You may wonder why not use B as the calling object. There are three reasons why the flag was kept. First, it is a part of the proposed iBLAS standard. Second, it allows B to be the result of a calculation; for example

`A.mm((M+N), C, false, false, false)` $\Leftrightarrow (M + N) * A + C$

Finally, the subclasses of `IntervalMatrix<I>` have specialized memory structures. This allows a calculation to take advantage of knowing the memory structure of the calling object. This means that for $B * A + C$, where A is, for example, a banded matrix, it is faster to call `A.mm(B, ...)` than to call `B.mm(A, ...)`.

10.4.6 Structured Matrices

There are three basic categories of structured matrices in iBLAS. First is the general banded matrix. This is followed by triangular and symmetric matrices. Figure 10.1 presents the mathematical relationships of different kinds of interval matrices involved in iBLAS. It provides the base for the class hierarchy implemented in this toolbox.

The following derived classes from `IntervalMatrix<I>` are included in the library:

- `TriIntervalMatrix`
- `SymIntervalMatrix`
- `BandedIntervalMatrix`
- `TriBandedIntervalMatrix`
- `SymBandedIntervalMatrix`

Only differences between these specializations and `IntervalMatrix<I>` are discussed. All of the specialized matrices are square. Hence, the constructors of structured matrices take only a single dimension parameter instead of

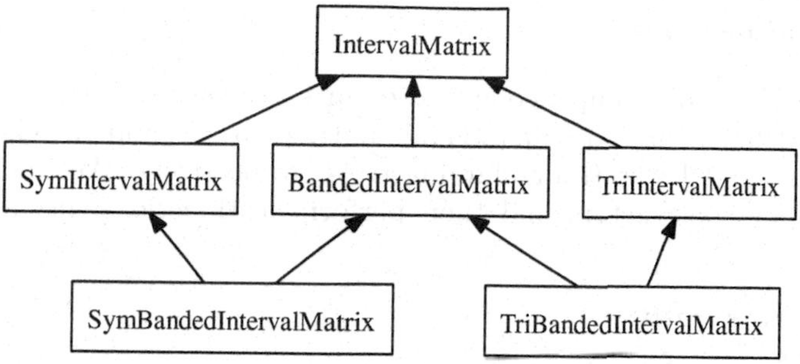

Fig. 10.1. Mathematical relationship of interval matrices in iBLAS.

two. Single inheritance is used in this library since it greatly simplifies both the coding and understanding of the library.

The class `BandedIntervalMatrix<I>` implements banded matrices. When constructing a banded matrix, the user will need to supply the number of subdiagonals and superdiagonals. Triangular interval matrices are symbolized by the class `TriIntervalMatrix<I>`, and symmetric interval matrices are represented by the class `SymIntervalMatrix<I>`. The derived classes `TriBandedIntervalMatrix<I>` and `SymBandedIntervalMatrix<I>` can be used to effectively process matrices that are triangular and banded, or symmetric and banded. Both classes inherit and use the memory structure from the class `BandedIntervalMatrix<I>`. Of these two classes, the `TriBandedIntervalMatrix<I>` class only cares about the number of subdiagonals for lower triangular and the number of super-diagonals for upper triangular. The `SymBandedIntervalMatrix<I>` uses the same number for both of the sub-diagonals and super-diagonals. This emphasis on the diagonals is the reason why the two classes inherit from `BandedIntervalMatrix<I>`. However, the operators and function names involve these specialized matrices are the same as those in the general `IntervalMatrix`. For more information about these derived classes, users may refer to the documentation for this toolbox.

10.4.7 Extension

The OOP design of this toolbox provides a great deal of flexibility for a user to extend the library with specialization. To implement a new matrix type (e.g., `UserDefinedIntervalMatrix`), that maintains the interoperability with the rest of this toolbox, one must meet the following requirements. First, the user-defined type must inherit, either directly or indirectly, from `IntervalMatrix<I>`. Second, if the user-defined class uses a different storage mechanism than its parent class, then the user must overload the `getAtNC()` and `setAtNC()` methods. Finally, the methods `getRows()` and `getCols()` must function correctly. If the user follows the three steps outlined here, then the new class will seamlessly interoperate with the rest of this toolbox.

10.4.8 Sample Code

Here is a segment of example code that defines and accesses interval matrices. The code defines three 3×3 interval matrices a, b, and c, where b and c are upper and lower triangular matrices, respectively. We initiate them with different constructors and then perform arithmetic, logic, and utility operations.

```
#include <iostream>
#include <iomanip>
#include <IntBox.hpp>

using namespace std;
using namespace intbox;

int main() {
  IntervalMatrix<> a(3,3);
  TriIntervalMatrix<> b(3, true ), c(3, false);

  b[0][0] =  default_interval( -0.1, 0.1 );
  b[1][0] = -3; b[1][1] =  2;
  b[2][0] =  5; b[2][1] = -1; b[2][2] = 5;


  c[0][0] = 0; c[0][1] = -3; c[0][2] =  5;
               c[1][1] =  2; c[1][2] = -1;
                             c[2][2] =  5;
  if ( b.enc( c.transpose() ) )
    cout << "B enclose C^T\n" << endl;

  cout << setprecision( 3 ) << setiosflags( ios::fixed );
  a = b * c * default_interval( .5 );
  cout << "A = B * C * [0.5 ; 0.5]\n" << a << endl;

  return 0;
}
```

The above code segment outputs:

```
B enclose C^T

A = B * C * [0.5 ; 0.5]
|  [0.000 ; 0.000]  [-0.150 ; 0.150]  [-0.250 ; 0.250]  |
|  [0.000 ; 0.000]  [6.500 ; 6.500]  [-8.500 ; -8.500]  |
|  [0.000 ; 0.000]  [-8.500 ; -8.500]  [25.500 ; 25.500]  |
```

In the output, we use a semicolon to separate the lower and upper bounds of an interval as suggested by the proposed C++ interval standard library.

10.5 The `IntervalVector` Class

This software toolbox uses a single one-dimensional interval array to store all of the elements in an interval vector. Unlike matrices, we do not consider any specializations of its structure. Hence, only one class is needed in the implementation, namely `IntervalVector<I>`.

10.5.1 Creation and Assignment

There are three constructors in the class. The most commonly used one simply takes the length of the vector. Another constructor creates a vector from two arrays, where the first array is the lower bound and the second array is the upper bound of an interval vector. For robustness, an element in the first array does not necessarily have to be less than the corresponding element of the second. If it happens, the two elements will simply be reversed. If needed, the user should verify if the lower bound is actually less than or equal to the upper bound for each interval in the input data. The size of vector is set by the constructor and is fixed. The size of a vector can be retrieved using the `getDimension()` method. The default constructor creates a vector of size zero, with no underlying array. It is best used for catching returned values similar to the default constructor of `IntervalMatrix<I>`.

As mentioned earlier, the size of a vector cannot be resized. As with matrices, there is one exception to this rule, and that is assignment. All of the principles dealing with matrix assignment are also applied to vector assignment. The elements of a vector can be accessed the same ways that matrix elements can be accessed.

10.5.2 Overloaded Operators and Boolean Methods

As previously mentioned, this library also overloads the arithmetic operators for elementwise arithmetic operations of interval vectors such as addition and multiplication. These overloaded operators are `+` (addition), `-` (subtraction), and `*` (multiplication). Each of these operators, whenever appropriate, is valid for vector-vector, vector-interval, and interval-vector operations. As in matrices, each of these operators has an associated method that performs the operation (i.e., `add()`, `sub()`, and `mult()`). It is recommended that these methods be used directly if the user is utilizing pointers.

It can be useful to compare two interval vectors. With this in mind we implemented several Boolean methods for interval vectors in this toolbox. First, the == (equal) and != (not equal) operators have been overloaded. Two vectors are considered equal if each corresponding element in both vectors is equal. The method isEmpty() will return true if the vector contains an empty element. Furthermore, as with matrices, the Boolean methods enc(), interior(), and disj() have also been implemented.

10.5.3 iBLAS Methods for Interval Vectors

The IntervalVector<I> class uses the same approach to implementing the iBLAS methods as discussed in the previous section. The parameters of iBLAS methods of this class start with other vectors followed by any scalars. For example, below is the function header for the axpby() method, which computes the vector $\alpha * \mathbf{x} + \beta * \mathbf{y}$.

```
IntervalVector<I>& axpby( IntervalVector<I>& y,//A Vector
                          const I& alpha,      //A Scaler
                          const I& beta,       //A Scaler
                          IntervalVector<I>& result_vec )
```

The calling object takes the place of the **x** in the formula. This is a general rule for iBLAS functions. The parameters are structured based on the formulas given above. The parameter result_vec is an *optional* parameter that always appears last in every method. It can be used to store the return value. If a vector for result_vec is specified, then that vector will be used instead of a new or recycled one. Furthermore, if the result_vec parameter is specified, then a reference to result_vec is returned.

10.5.4 Sample Code

Here is a segment of example code that defines and accesses interval vectors and matrices. The code defines a 3×3 interval matrix a and two interval vectors b and c.

```
#include <iostream>
#include <iomanip>
#include <IntBox.hpp>

using namespace std;
using namespace intbox;

int main() {
  IntervalMatrix<> a(3,3), result;
  IntervalVector<> x(3), y(3);
```

```
  a[0][0] =  1; a[0][1] = -7; a[0][2] = 0;
  a[1][0] = -3; a[1][1] =  2; a[1][2] = 6;
  a[2][0] =  5; a[2][1] = -1; a[2][2] = 5;

  x[0] = default_interval( -5.3, -0.9 );
  x[1] = default_interval( -1.0e-3, 0 );
  x[2] = 16;

  y = x * default_interval( -10, 1.25 );

  x = x.axpby( y, 1, default_interval( .5, .6 ) );
  result = a.r( x, y ); //Rank one update of matrix

  cout << "result=(x+[-10;1.25]*x*[.5;.6])+[-10;1.25]*x*a\n"
       << result << endl;

  return 0;
}
```

The above code segment outputs:

```
result=(x+[-10;1.25]*x*[.5;.6])+[-10;1.25]*x*a
|  [-490.575 ; 1638.7] [-7.09275 ; -6.691] [-4944 ; 1484]  |
|  [-3.09275 ; -2.682] [1.99998 ; 2.00006] [5.04 ; 6.28]  |
|  [-4235 ; 1489] [-1.8 ; -0.72] [-4475 ; 12805]  |
```

10.6 Obtain, Install, and Use the Toolbox

10.6.1 Obtain the Toolbox

Through `www.cs.uca.edu/interval`, the interval computations homepage at the University of Central Arkansas, this package can be downloaded online for free. The toolbox can be obtained in two forms. First, you can download a binary installer for Windows. Alternatively, you can download a gziped tarball, which is suitable for installation on Unix-like operation systems. Both packages contain the complete documentation, source code, tests, sample application programs, the license, and a README file. The documentation offers lengthy descriptions of all the files, classes, class members, and the `intbox` namespace. It is written in HTML format and organized in the `doc` directory of the package.

10.6.2 Install and Test the Toolbox

For easy installation and testing of this library, a single makefile is provided for machines with a `GCC` compiler, and a binary installer is provided for installing on machines with Microsoft Visual Studio .NET 2003. If the user already has Boost installed on the computer, then neither the binary installer nor the make file will install the Boost interval library; that is, it will not overwrite the user's installation. Detailed installation instructions as well as information on porting the toolbox to unsupported platforms or compilers is provided in the `Readme.html` file. The package is licensed under the Boost Software License version 1.0.

We have installed and tested this library on various processors, operating systems, and compilers. The processors include an Intel Core Duo processor, an Intel Pentium 4M, Sun SPARC IV, and an AMD-64 processor. The list of operating systems used to test the library consists of Windows XP using both Microsoft Visual Studio .NET 2003 and CygWin, Solaris 10 using GCC, and, finally, Ubuntu Linux 7.10 also using GCC. The compilers used include Microsoft Visual Studio .NET 2003, GCC 3.4.3-4, and GCC 4.1.2. Written in standard C++, this toolbox should work for any computer with a modern C++ compiler. For those who may experience installation problems on other platforms, please refer to the `Readme.html` file for further information.

10.6.3 Using the Package

To use the library in applications, the user need only do two things. First, one should include the header file `<IntBox.hpp>`; this is the only file needed to access the library. Second, one should make the appropriate `using` declarations, since all the classes are in the `intbox` namespace.

Here is a sample program. It defines and manipulates intervals, interval matrices, and interval vectors.

```
#include <iostream>
#include <iomanip>
#include <IntBox.hpp>

using namespace std;
using namespace intbox;

int main() {
  default_interval a = 2;
  IntervalMatrix<> A( 2, 2 ), M, N;
  IntervalVector<> x( 2 ), y( 2 ), r;

  A[0][0] = default_interval( -1.1, 5 );
  A[0][1] = default_interval( 2, 3.3 );
  A[1][0] = default_interval( -1, 3.3 );
  A[1][1] = 5.7;

  x[0] = 4;  x[1] = default_interval( 4.9, 5.1 );
```

```
  y[0] = 10; y[1] = default_interval( 10.95 );

  try {
    M = A.mid();
    M = A - M;
    r = (a*A*M).mv( x, y );

    cout << setprecision(17) << showpoint;
    cout << "A - A.mid() =\n" << M << endl;
    cout << "aAMx + y =\n" << r << endl;

  } catch ( INTERVAL_EXCEPTION ie ) {
    cout << ie << endl;
  }
  return 0;
}
```

The above sample program can be compiled using Microsoft Visual Studio .NET 2003 using the standard Win32 Console project type. Using GCC, the sample program can be compiled on the command line as follows:

```
gcc -o sample sample.cpp
```

The above code segment outputs the following. To fit the output on the page, we omitted seven or more digits in the first output matrix.

```
A - A.mid() =
|  [-3.050...003 ; 3.050..003] [-0.649...991 ; 0.649...991] |
|  [-2.149...999 ; 2.149..999] [0.000...000 ; 0.000...000]  |

aAMx + y =
{ [-201.91000000000003 ; 221.91000000000003]
  [-189.48900000000006 ; 211.38900000000007] }
```

The documentation of this package provides more detailed information for using this toolbox.

10.7 Conclusions

This interval toolbox provides a freely available, well tested, portable, and robust coding tool for interval software development, with extensive documentation. It can be easily embedded into any standard C++ environment. With an object-oriented design and implementation in C++, this toolbox is easy to use, even for beginners with little knowledge of interval computing.

Acknowledgment: The authors would like to express their gratitude to Dr. R. Pozo, Dr. R. B. Kearfott, Dr. P. Young, and Dr. A. Goldsztejn for their insightful comments which helped us to improve the package. The authors also thank the developers of the Boost interval library. This work is also partially supported by the U.S. National Science Foundation under grants CISE/CCF-0202042 and CISE/CCF-0727798.

References

1. Blackford, G., Demmel, J., Dongarra, J., E.A.: Basic linear algebra subprograms technical (BLAST) forum standard. High Performance Computing Applications 16(1-2), 1–199 (2001). `http://www.netlib.org/blas/blast-forum/`
2. Blackford, G., Demmel, J., Dongarra, J., Duff, I., Hammarling, S., E.A.: An updated set of basic linear algebra subprograms (BLAS). ACM Transactions on Mathematical Software 2(28) (2002)
3. Brönnimann, H., Melquiond, G., Pion, S.: A proposal to add interval arithmetic to the C++ Standard Library. Technical proposal N1843-05–0103, CIS, Brooklyn Polytechnic University, Brooklyn, NY (2005). URL `http://boost.org/libs/numeric/interval/doc/interval.htm`
4. Melquiond, G., Pion, S., Brönnimann, H.: Boost C++ libraries: Interval arithmetic library (2002). URL: `http://www.boost.org/libs/numeric/interval/doc/interval.htm`
5. Nooner, M., Hu, C.: A computational environment for interval matrices. In: R.L. Muhanna, R.L. Mullen (eds.) Proceedings of a workshop on Reliable Engineering Computing, pp. 65–74. Georgia Tech. University, Savanna, GA (2006). `http://www.gtsav.gatech.edu/workshop/rec06/proceedings.html`
6. Pryce, J.D., Corliss, G.F.: Interval arithmetic with containment sets. Computing 78(3), 251–276 (2006). DOI `http://dx.doi.org/10.1007/s00607-006-0180-4`

Index